AF322623

A Handbook on Food Packaging

THE AUTHOR

Dr. P. Jacob John, presently working as Professor and Head of the Department of Processing Technology, College of Horticulture, Kerala Agricultural University, Vellanikkara, Thrissur. Kerala, is an Agricultural Graduate from Kerala Agricultural University and did his Post graduation and Doctoral Degree in Food Technology from CFTRI Mysore. He was the recipient of ICAR Senior Research Fellowship for his PhD programme. He has more than 30 years of experience in teaching, research and extension activities. As the Head of the Department he has guided more than 33 postgraduate students and personally guided over 10 students in the field of processing and post-harvest technology. He is a member of the Board of Studies of Kerala Agricultural University, Calicut University, and Mahatma Gandhi University, Programme in charge of IGNOU study centre for diploma course in fruit and vegetable value addition. He is the Project Co-coordinator of post-harvest technology research programme of the University. During 1999 he was the President of the Association of Food Scientists and Technologists (AFST, India), Thrissur Chapter. He has attended an International Training Programme in R&D on Post-Harvest Practices held at Agricultural Research Organization of Israel, as an invited guest of the Government of Israel.

He has to his credit several national and international papers, popular articles and patents. He has also wirtten a book "A Handbook on Postharvest Management of Fruits and Vegetables" Published by Daya Publishing House, New Delhi.

A Handbook on Food Packaging

By
P. Jacob John
Professor and Head,
Department of Processing Technology,
College of Horticulture,
Kerala Agricultural University,
Vellanikkara, Thrissur, Kerala

2022

Daya Publishing House®

A Division of

Astral International Pvt. Ltd.
New Delhi - 110 002

© 2017, AUTHOR
Reprint 2022

ISBN: 9788170359289

Publisher's note:
Every possible effort has been made to ensure that the information contained in this book is accurate at the time of going to press, and the publisher and author cannot accept responsibility for any errors or omissions, however caused. No responsibility for loss or damage occasioned to any person acting, or refraining from action, as a result of the material in this publication can be accepted by the editor, the publisher or the author. The Publisher is not associated with any product or vendor mentioned in the book. The contents of this work are intended to further general scientific research, understanding and discussion only. Readers should consult with a specialist where appropriate.

Every effort has been made to trace the owners of copyright material used in this book, if any. The author and the publisher will be grateful for any omission brought to their notice for acknowledgement in the future editions of the book.

All Rights reserved under International Copyright Conventions. No part of this publication may be reproduced, stored in a retrieval system, or transmitted in any form or by any means, electronic, mechanical, photocopying, recording or otherwise without the prior written consent of the publisher and the copyright owner.

Published by : **Daya Publishing House**®
A Division of
Astral International Pvt. Ltd.
– ISO 9001:2015 Certified Company –
4736/23, Ansari Road, Darya Ganj
New Delhi-110 002
Ph. 011-43549197, 23278134
E-mail: info@astralint.com
Website: www.astralint.com

Foreword

Packaging forms an integral part of food manufacturers, providing the essential link between the processors and consumers. Small family size and increased number of working adults, industrial growth and urbanization, living and eating habits are among the changes that has brought a drastic demand for processed and convenient food which are mostly depending on the packaging systems and materials.

There is an urgent need to look into the aspect of modern packaging which forms an integral part of food processing. Food processing is again an emerging field which is now taken up as one of the important subject for undergraduate and postgraduate students in Agriculture, Horticulture, Dairy Science, Food Science and Technology, Agriculture Engineering and so on. In order to facilitate such students, faculty and workers engaged in this field, with the knowledge on food packaging, an attempt has been made in this hand book to compile and present the informations available on food packaging into a handbook form. I am sure that *"A Handbook on Food Packaging"* will lead to more scientific and methodological approach to food packaging with the idea in protection, preservation, consumer attraction, safety and hygiene in food processing and packaging.

I hope this book will serve as a ready-reckoner for the food processors, students and researchers in the field of food packaging.

Prof. (Dr.) P.K Ashokan

Director,
Academic and P.G. Studies
Kerala Agricultural University
Thrissur, Kerala

Preface

Packaging in India until recent years was considered a luxury. This perception was mainly due to the fact that most materials were available for consumption without any packaging and packed material of the same commodity were more expensive. As the consumer awareness and the standard of living increased the need for packaged commodities also increased. Packaging supported by sophisticated and modern technologies are geared to develop the requisite packaging media and systems envisaging a dramatic turn in the marketing of food and non-food products.

Packaging forms an integral part of food manufacture, providing the essential link between the processors and consumers. Small family size, increased number of working adults, industrial growth and urbanization, living and eating habits are among the changes that has brought a drastic demand for processed and convenient food which are mostly depending on the packaging systems and materials.

Food packaging is defined as a means or systems by which a fresh produce or processed product will reach from production centre to the ultimate consumer in safe, sound and hygienic condition. It is a combination of art, science and technology. It forms a part of production, storage, handling, distribution, retailing and end use.

Considering the importance of packaging, it has become a subject of learning in Agriculture, Horticulture, Dairy Science, Food Science and Technology, Agricultural Engineering etc. Informations on food packaging are available through various publications of Central Food Technological Research Institute (CFTRI) Mysore, Indian Institute on Packaging Technology, Mumbai. A compilation of all these informations including the relevant informations gathered from net and journals have prompted me to develop "A Handbook on Food Packaging". I acknowledge the above publications to form a basis for this book. I also feel this hand book will defiantly provide some basic ideas for the students, faculty and research workers in the field of food processing and packaging as a ready reckoner.

P. Jacob John

Contents

Chapter 1
Introduction and History

Economic liberalization in India has opened the doors of massive expansion in investment and production in the entire spectrum of industry along side of this substantial deepening of the industrial structure, the age of high mass consumption also seems to prospect. India being identified as one of the largest market for consumption goods in Asia, more than 200 million middle class people in India has clearly sent the message to the whole world that their appetite for consumer good is enormous by any standard. What does all this mean to marketing and packaging which is one of the crucial imperative for the competitive marketing?

Walk into any shopping arcade, the array of consumer goods ranging from provisions of everyday use, cosmetics, specialty products, and convenient foods, ready to cook or serve foods, detergents and what not proclaiming that the days of the sellers market have virtually disappeared and the consumers market is the order of the day. Manufacturers have to come to terms with the reality of consumer power and that explains how aggressive marketing with sustained brand promotion strategies has become the order of the day. In this rough and tumble of competitive marketing, packaging has become in its own right winning strategy. Packaging design and its material content have assumed a degree of importance as never before.

Packaging in India until recent years was considered a luxury. This perception was mainly due to the fact that most materials were available for consumption without any packaging and the packed materials of the same commodity were expensive. As the consumer awareness and the standard of living increased the need for packaged commodities also increased.

In India, the packaging industry has to serve two classes of society, one aware of the importance of the packaging and other not so. The later however, constitutes the majority of the population but nevertheless needs the product in packed form, therefore Indian packaging industries forced to serve all the classes of the society, create and promote packaging awareness among the masses and to meet export packaging requirements. Packaging supported by sophisticated and modern technologies are geared to develop the requisite packaging media and systems envisaging a dramatic turn in the marketing of food and non-food products.

Packaging forms an integral part of food manufacture providing the essential link between the processors and the consumers. Small family sizes and increased number of working adults, industrial growth and urbanization, living and eating habits are among the changes that has brought a drastic demand for processed and convenient food which are mostly depending on the packaging systems and materials.

Food packaging is defined as a means or system by which a fresh produce or processed product will reach from the production centre to the ultimate consumer in safe hygienic and sound condition at an affordable price. It is a combination of science, art and technology. It forms a part of production, storage, handling, distribution, retailing and end use. It is also considered as the last output of production and the first input of marketing

Following chapters of this hand book briefly explains the packaging materials, their characteristics and different methods of packaging to suit each product for maintaining quality and life.

History

The subject "food packaging" has been introduced to modern technology recently but this was existed on earth without having any proper attention. Most of the agricultural produces like coconut, orange, banana etc having a husk, skin or peel are available to use in packaged form having good hygiene. This type of packaging is considered to be natural packaging. Further, the use of dry leaves for wrapping meat, animal skin for making storage bags were in vogue since long.

One hundred years ago there was little use for packaging in food industries. Products were sold at small local stores or markets, often from bulk containers to the containers brought or provided by the consumer for his/her purchase. Butter for people was sold from large blocks wrapped in a piece of paper for the consumer to take home. Liquids were kept in bulk containers and served out into the customer's container.

Packaging dates back when people first started moving from place to place. Originally skins, leaves and barks of plants were used for food transport. The Greeks and Roman times saw the use of pottery and the start of glass. In the 1400 s wood were used to make containers and after 1850 paper from wood and the use of paper and glass started. Mesolithic humans used baskets for transportation whereas Neolithic humans started using metal containers and pottery jars were used to protect against rodents; in 1550 BC glass making was an important industry in Egypt, but glass remained expensive until the eighteenth and nineteenth centuries. Tin coating on iron plates became possible in AD 1250. In 1825 aluminium was first extracted. Napoleon Bonaparte was the man behind for the evolution of metal containers known then as canisters for canning of food products during the war time.

In spite of all the efforts taken to produce high grade processed products, unless they are delivered in fresh, sound and convenient form to the consumer, they are likely to be rejected, thus causing wastage. This loss can be minimized to great extent by adequate protective packaging to withstand the hazards of climatic changes, transportation and handling hazards.

The spurt in the growth of the processed food industry has a bearing on the development of improved packaging materials; and in today's society packaging is pervasive and essential. Without proper packaging handling would be messy, inefficient and costly exercise. The highly sophisticated packaging industries of today which characterize modern society are far removed from the simple packaging activities of earlier times. Food packaging is an area sooner or later, every practicing food scientists and technologists become involved and the challenges before them is to design and develop functional packaging to suit each product and its purpose at minimum cost. Therefore, an attempt has been made to gather and provide certain available information on food packaging in a hand book form.

References

Anon (1995) The vital "P" in Marketing. in the daily "Hindu" dated 9/2/95.

Abjith kar (2005) Packaging for physical distribution and transportation. In the lecture schedule for the DVAPFV course of IGNOU, New Delhi.

Dordi, M.C.(1995) Food packaging trends. Journal on Food Science and Technology, AFSTI Mysore.

Durscoll, R.H.and Petereson, J.L (1999) Packaging and food preservation. In: Handbook on Food Preservation Ed. Rahman M.S., Marcel Dekker. Inc. New York.

Saha, N.C. (2005) Food Packaging Needs and Importance. In the lecture schedule for the DVAPFV course of IGNOU, New Delhi.

Chapter 2
Functions of Packaging

The functions of packaging are performed by three P's *i.e.*, preservation, protection and presentation. The above P's are applicable to consumer products. Packaging serves two basic objectives *i.e.*, marketing and logistic functions:

Marketing Functions

☆ It provides information about the product to consumers.

☆ It promotes the product with attractive graphics and printing

☆ It acts as a medium of communication.

☆ It acts as a silent salesman and the final interface between the company and the consumers.

Logistic Functions

☆ Containment

☆ Protection

☆ Apportionment

☆ Unitization

☆ Convenience

☆ Communication

Major function is protection of the product from various external agencies which are trying to destabilize the product soon after it is processed or manufactured.

These agencies are illustrated below.

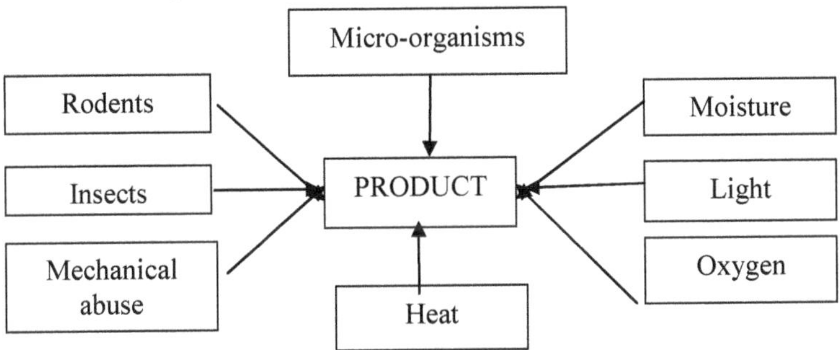

The order of preference of these agencies against a product will depend upon the nature of the product. For example in case of spray dried powders moisture will be the number one enemy, where as for the fatty foods, light may be the one and for fragile materials mechanical abuse may be the one external agency causing damage.

The package function is closely associated with many other functions in an organization and are inter-related and is illustrated below.

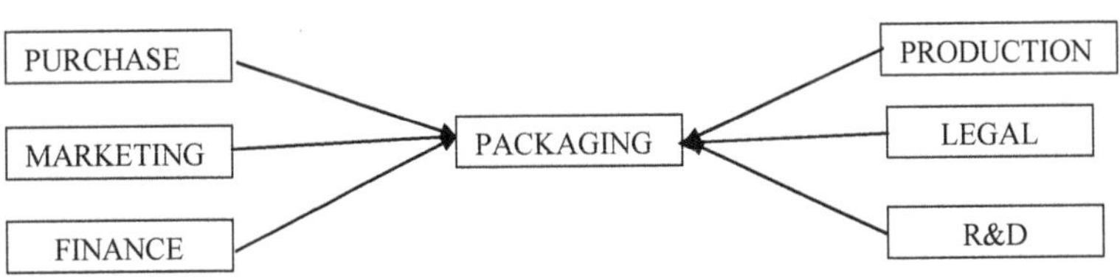

This shows the fact that packaging department is considered to be the nucleus in any fast moving consumer goods manufacturing company. The package should be designed to provide efficient storage; the design is influenced by

Product, (its nature and adaptability).

- ☆ Price or cost
- ☆ Protective level needed to suit the product
- ☆ Handling ability
- ☆ Marketing trends
- ☆ Reusability and recyclability of the material.

Therefore for a design consideration we should broadly have the following knowledge.

- ☆ Knowledge about the product
- ☆ Knowledge about the transportation hazards
- ☆ Knowledge about the market
- ☆ Knowledge about the packaging materials, machinery and labour.

Knowledge about the Product

The following informations needed to be available regarding the product prior to design consideration.

- ☆ Material from which the product is made and the manner in which these can deteriorate (Whether plant source, animal source, marine source etc)
- ☆ Shape and size of the product
- ☆ Weight and bulk density
- ☆ Its strength and weakness (which part is having load bearing property, *e.g.,* eggs if placed vertically

have the load bearing ability than its horizontal position).

Moisture criticality of the product whether it is absorbing or desorbing in nature (*e.g.* powder are highly hygroscopic and hence absorbing type while fresh fruits and vegetables are rich in moisture hence desorbing type).

Compatibilities with the packaging materials and vice-versa.

1. Physical state of the product whether liquid, solid, powder, granules, capsules, viscous fluids, corrosive, volatile, sticky, fragile, toxic, perishable, abrasive etc.

2. How the product can be damaged by mechanical shock, or vibrations, abrasions, crushing due to temperature or moisture, oxygen or light or due to insects or rodents or tainting due to foreign odour etc.

3. Biological hazards like ants, rodents, termites and insects

Knowlege about the Transporation Methods and its Related Hazards

Damage in transportation is one of the oldest problems in packaging. The hazard that package may not be adapted. Therefore protection is requested against the average hazard encountered. The distribution system is mainly divided into two.

☆ Private transport system

☆ Public transport system

Private system is the one managed by the manufacture or the distributor himself for the product; where the level of hazards encountered to his product is minimized due to adequate protection, he can provide while in transit and public is through rail, road, ship or air where the manufacturer or distributors do not have direct control over the system. The distribution pattern of a consumer good is given below.

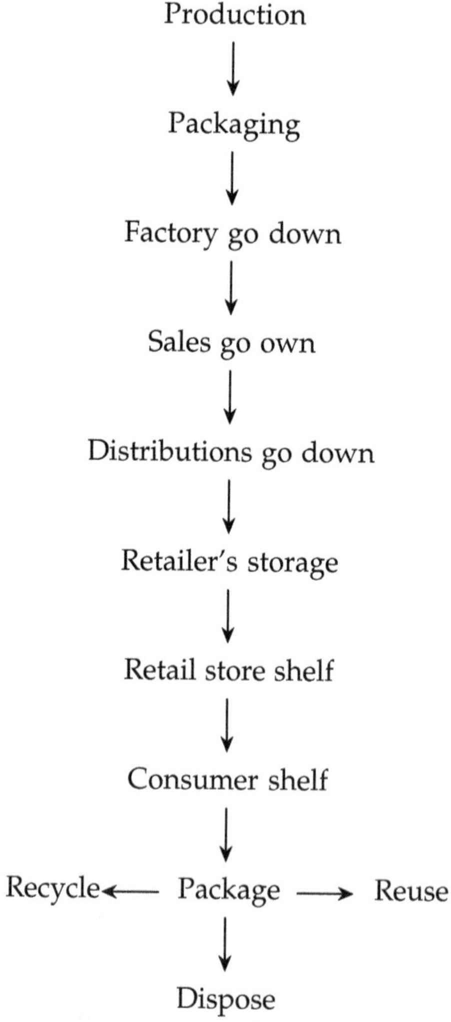

The distribution hazards encountered and which needs to be taken care to decide the packaging material depends upon the mode of transportation, method of handling and storage. Various transportation methods and their hazards are.

Bullock or Horse Drawn Cart

Mainly used for short distance transportation.

Stacking height up to 5 ft.in the cart

Drop height up to 5 ft. from the road level to the cart

Bumping due to rough road surface. Directional placement of packages may not be possible.

In spite of all these hazards more than 50 per cent of goods in the rural areas still resort to this mode of transportation due to country roads and less expensive transport means.

Road Transport

Mainly used for long and short distances transportation. Package dimensions should suit the body dimensions of the vehicle for maximum space utilization.

Delays in journey during rainy season, freight rates generally higher than rail.

Lower standard packages may be accepted than in rail.

Hazards

Stacking height up to 7 ft in the truck

Drop height up to 4 ft from the road level to the truck.

Puncturing of fibre board boxes by protruding bolts of the sides

Bumping due to irregular road surface

Rail Transport

Issues

Transshipment is necessary

Pilferage is a major problem

Proper handling instructions are essential

Packages have to be sent on railway risk

Packages are to confirm the rules and regulations of the railway

No problem for inter state transport

Less interruption of journey in rainy season

Hazards

Stacking height up to 8 ft in the wagon.

Vibrations due to rail joints and track conditions

Shunting shocks

Very high temperature in steel wagons during summer season.

Sea Transport

Packages carried in ship and docks normally faces journey preceded and followed by other modes of transport.

Require proper handling instructions in different languages and figures.

Freight rates are always by cubic volume.

Hazards

Stacking height up to 15 ft

Manual or machine handling

Very high relative humidity in the holds

Salt spray water in the ducts or on the docks lead to rusting due to the salt content in the sea water

Vibration due to engine propeller

Swaying due to waves

Air Transport

High freight charges hence lighter packages, has to be used

Limitations on size and weight of packages

Less journey time

Better handling

Hazards

High frequency vibrations of the engine.

Low temperature and low pressure when flying at higher altitudes

Transportation Cost

The transportation cost would depend on many factors like methods of transportation, efficiency of transportation, input cost for transportation, allowable time limit for transportation, losses during transportation, cost of the produce that is being transported and other associated costs. Therefore method of transportation must be selected with aim to minimize losses and optimize the profit.

Knowledge about the Marketing Systems

The following informations are to be gathered while designing a package for a product. The existing market competition for the product manufactured by others and the types of packaging materials and graphics already with that product.

- ☆ Quantities sold and in which unit pack size.
- ☆ Price of these products.
- ☆ How the product is sold, through self service, departmental stores, mail order, door to door or through super markets.
- ☆ Details about the target group (Consumers) age group, gender, social level, region, and income level.

Knowledge about the Packaging Materials, Machines, Laws and Regulations

These aspects are discussed in detail in the entire text under various chapters.

References

Mahadeviah, M. (1985) Recent developments in Food Packaging materials, Indian Food Industry.

Saha, N.C. (2005) Food Packaging needs and importance. In the lecture schedule for the DVAPFV course of IGNOU, New Delhi.

Sathis, H.S (1995) Packaging for physical distribution in the book Profile on Food Packaging published from CFTRI.

Chapter 3
Classification of Packages and Packaging Materials and Techniques

Based on the availability of the materials it is classified as traditional or natural and fabricated or modern packaging materials. In the past only two important functions were met *viz;*

☆ To contain the material.

☆ To carry the material.

Whereas in the fabricated once, apart from the above two additional functions like protection and sales are also incorporated. Traditional or natural ones were selected or used depending on the availability and convenience of such materials from in and around the place of production or sales of the items intended for packaging. Some of the examples of natural materials used in the earlier days are as follows:

1. Bamboo based containers like baskets.
2. Mat based containers made from fibre or leaves of plants.
3. Leather based containers from the hides of animals.
4. Clay based containers like pottery.
5. Jute based *e.g.* gunny bags.
6. Textile based-cloth bags.
7. Wood based- boxes, crates,
8. Plant parts *e.g.* fruits of bottle gourd.
9. Large leaves of teak, arecanut leaf sheaths etc.

Leaf Based

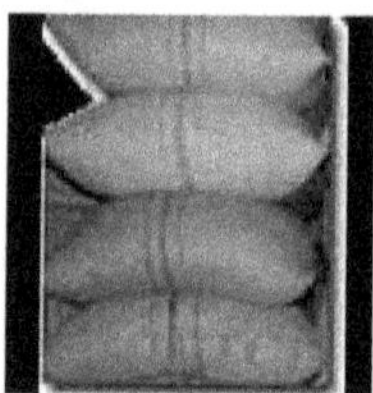

Jute Based

Clay Based

Wood Based

Fabricated or Modern Packaging Materials

Classification is again divided into rigid and flexible materials.

Rigid Containers

Rigid packaging materials are still continue to hold its importance in spite of the astonishing growth of flexible packaging materials, owing to its load bearing property and adequate protection for soft materials as well as fluids. Major rigid containers included are:

- ☆ Metal containers like drums, barrels, tin, Aluminium and tin free steal (TFS).
- ☆ Glass containers like bottles, jars, ampules etc
- ☆ Wooden containers like boxes and crates.
- ☆ Plastic containers like bottles, barrels, Jericans, drums, plastic crates.
- ☆ Paper drums.
- ☆ Plywood containers.

Semi Rigid Containers

- ☆ Aluminium collapsible tube.
- ☆ Plastic collapsible tube.
- ☆ Composite container.
- ☆ Paper based cartons.

Flexible Containers

The emergence of plastic as a versatile substitute for metal in an enormous range of applications explains much of flexible packaging. It is not plastics alone lend themselves to use for flexible packaging. Metal foils and various types of paper also form different combinations of these materials as laminates has really revolutionized the packaging industry as flexible containers. In the process of lamination the weakness of one material will be overcome by the strength of the other, thus the final combination will be strong enough to protect the content from all those external agents involved in destabilizing the product.

Classification of Packaging Components

Unit Pack or Primary Pack

It is considered as a package containing a single unit *e.g.* Toffee in a wrapper, unit package should protect the product against the deterioration of quality; naturally the material selected should have adequate functional properties to serve such functions. This package also serve as a consumer pack or retail package, therefore should have eye appeal, easy operability, easy to carry or handle and customer friendly. This is also called primary pack because it is the prime material which is coming in direct contact with the product. A primary pack can be a consumer pack when there is no secondary pack needed to protect the primary pack.

Secondary Pack

A secondary package is a packaging material used as a protective package for the primary package for *e.g.* In the case of butter paper carton will be the secondary package and butter paper will be the primary package and this together will form a consumer package where graphics, authenticity and mandatory requirements needed is to be printed on the secondary package.

Intermediate Pack

Unit packs or consumer packs further packed into a retailer container to satisfy the marketing requirements are called intermediate pack. For *e.g.* 10 unit packs of 100gram butter packed in to a paper board carton to facilitate handling and enhance display value are called intermediate package, this can also be called a retailer pack.

Shipping Container or Tertiary Package

The intermediate packages are further packed into corrugated fibre board boxes (CFB) or wooden boxes or jute bags or HDPE woven sacks to transport packages. These packages provide protection to the contents from journey hazards like shocks, vibrations, drops, climatic changes etc. during handling, storage and transportation. Some of the most common shipping containers are given below.

Shipping Container

Wooden Containers

Materials used for manufacture of wooden boxes include natural wood and industrially manufactured wood based sheet materials. "Poplar" is widely used for peeled boards. Poplar wood is light in weight and thus reduces the empty weight of the package. Pine is also used for peeled boards and more particularly for corner posts. Manufactured wood based sheet materials include plywood, hard board and particle board.

Wooden Shipping Container

Wooden containers are one of the earliest shipping container since wood was then available in plenty. But now wood containers are not encouraged as forests are depleting year by year. There is stiff competition from fibre board containers both in terms of cost and performance. Still wood containers are widely used for perishables and machineries.

The optimum moisture content for wood is between 12 to 18 per cent for the manufacture of the containers. For the domestic transportation of perishables it is most widely used and the requirement for such containers are (a) proper aeration is required for the dissipation of heat and exchange of gases.(b) needs high stacking strength and stability.(c) there should be no fungus and mould growth when stored at high relative humidity.(d) it should be dimensionally stable.

Corrugated Fibre Board Boxes (CFB)

The corrugated fibre board is made of paper board liners and corrugated medium. The structure of corrugated fibre board consists of a fluting medium running in a sinusoidal wave form between the two liners, thus separate the liners by a distance to obtain good stiffness. Unbleached virgin 'Kraft' is most appropriate for liner materials. It has high tearing resistance and a low rate of moisture absorption from air. Coating of the boxes with wax, bitumen or polyethylene will retard the penetration of moisture but are restricted in use because of higher costs and export trade restriction. The manufacturer's joint of corrugated fibre board boxes may be formed by means of staples, glue or tape.

In the case of packaging fruits and vegetable, ventilations for the dissipation of heat and exchange of gases are very much essential. Careful attention must be given to the number, size, shape and position of these holes without sacrificing the strength of the box.

The corrugated fibre boards are designated as per the number of piles, the flute height and number/30cms. Thus a board as 3 ply A flute has 3piles with every 30 cm length having 32 to 38 flutings of 4.5 to 4.7 mm height.

The adhesives used in the boxes have a bearing on the machineability and quality of boards. The duration of stacking and relative humidity that surrounds the package influences the estimation of safe packing load. Being a visco-elastic material the corrugated fibre board exhibits "creep" under static load conditions.

Other bulk transportation containers are:

Barrels and Drums

Metal and wooden barrels are commonly in use. The metal barrels include steel and aluminium. Aluminium barrels are mostly used for storing beer.

Drum   **Barrel**

Wooden barrels are made of staves bound together with hoops and may be "tight" or 'slack". The tight ones are used for storing heavy solids, semi solids and liquids. The wooden barrels are also used for storing and ageing of alcoholic liquors.

Drums

A drum is a cylindrical shipping container different from barrels in having straight sides and flat or bumped ends designed for storage and shipment as an un supported outer package that may be shipped without boxing or crating.

Drums can be either metal drums either aluminium or steel or fibre drums, ply wood drums and plastic drums. Metal drums are single wall with either double head, partial opening, with convex or flat full removable head construction. The capacity usually ranges from 12 to 110 gallons. The inner surfaces of the metal drums are coated with lacquers like oleo resinous types, phenolic resins, and vinyl and epoxy resins.

Fibre drums are made by fibre board of piles not less than 0.3mm thick. The capacity of these drums varies between 0.5 to 100gallons. The advantages are that they are not returnable and has good stackability, low tare weight and easy opening and closing features. Printing can improve their appearance.

Basic types of fibre drums are:

Plain drums–no moisture proofness.

Liquid tight drums–has rubber or plastic gaskets.

Water vapourproof drums-inside laminated with waxed paper or polyethylene.

Lined and coated drums- prevent direct contact between content and fibre board.

Fibre drums are generally used for shipping semisolids having a minimum viscosity of 5000cps. It is also used for liquid insecticides, fruit juices and other food stuffs. Ply wood drums are made of 3 ply veneer which laps and joined with staples. Ply wood drums are generally used to pack dry products and have an excellent weight to strength ratio. The polyethylene drums are rigid and self supporting. They are available between 5 to 55 gallon capacity. The advantage of these drums is flexibility, non toxicity, light weight, durability, high chemical resistance. They have high resistance to breakage.

Sacks

Sacks are generally made from flexible low cost materials like jute, textiles, paper and plastics. Paper sacks are made of two or more piles of Kraft paper which is pure sulphate paper having a substance of 70 gsm or more. Plastic sacks are generally made from polyvinyl chloride, polyethylene and polypropylene. The advantages are that they are better weather and impact resistant and weighs less than multi wall paper sacks. These sacks are made either through blowing or by extrusion.

Sacks of combinations of plastic and jute are also commercially available. These bags are highly suitable for packaging of dried whole spices, grains and other dry products. Fresh fruits and vegetables in bulk are generally transported from field to domestic markets in jute bags. But sacks do not provide support for the product against super imposed loads and also offers less resistance to impact loads and possibilities of sifting and spilling are common.

Bag-in-Box

In this system, the bag is supported on the outside by a rigid container made of paper board or corrugated board carton. It is generally used to fill a variety of liquid and dry products like tea, instant coffee, milk foods, baby foods,

glucose powder, biscuits, spices, aseptic and non-aseptic fruit juices, edible oil, ghee etc. The system is tamper proof and offers cost effectiveness. Depending upon the product wide range of plastic film laminates, co extruded barrier films, metallised film and aluminium foil can be used for inner construction

There are two methods of producing bag-in-box packages.

1. Lined carton system

2. Coated and laminated carton system

In the case of lined carton system, the inner liner is made from a suitable laminate such as LDPE/Paper/HDPE, Paper/foil/LDPE and polyester to give the required shelf life protection to the product. The outer carton is made of duplex board for protection against damages. The second method combines the carton forming/gluing operation with a lining feed mechanism.

Palletization

A pallet is a platform made to hold one or more boxes, bags, cartons etc in a group. The pallet is one of the simplest single devices for material handling. The advantages are:

☆ Reduced labeling requirements

☆ Better utilization of storage space because of higher stacking strength.

☆ Reduction in mechanical strains and damages.

Wooden Pallets with Modular Arrangements

☆ Reduction in the total distribution time.

☆ Better maintenance of produce quality.

A pallet can be made of wood, corrugated or honey comb paper board, plastic, or reinforced plastic or metal. The choice is based on the service conditions such as weight of load, climatic environment, durability requirement, local availability and cost.

There are two principles in the pallet assembly

(1) The Modular Principle

Here the packages are oriented in the same direction of the pallet; this method is called column stacking, in which the pallet loads will have rectangular cross section without internal spacing.

(2) The Two Way Principle

Here the packages in each tier form a pattern such that some packages are oriented lengthwise and others crosswise on the pallet. This type is called as interlocking stacking.

Pallet Dimensions

Standard dimension of 120x100 cm is the primary handling unit size for international trade and for countries like Germany, Switzerland, Austria etc the standard size is 120x 80 cm and for internal means of transportation the pallet size of 104x100 cm is mostly preferred. Pallets of 4 way entry is preferred for easy lifting and loading.

Inner Packaging Components

These components are those given to transport by way of providing the resistance of movement of contents during journey. Materials like thermocole, expanded poly ethylene, paper shavings, glass wool, wood shavings are used either to protect the product against shock hazards or as space fillers to prevent the jerks and jumping during journey, some of the examples are:

Cell Pack

Portioning inner packing method contributes to the stacking strength of the box. In medical pack; bottles are mostly separated by cell pack. Paper honey comb–each cell will be compartmentalized as honey comb like structures using paper separators and the space can be of different size and shape to suit the product. Moulded pulp trays– generally used for eggs, fruits and vegetables. Expanded polystyrene inserts–can be produced in short runs on reasonable costs as cushioning material.

Thermo formed PVC trays: Generally used for packaging produce in single layers.

Paper shavings or wood wool: To provide cushioning effect and to provide good protection on tight packing.

Plastic foam net: Generally for large size fruits and for glass bottles.

External Reinforcements

Shipping containers are further reinforced by means of either plastic or metal straps applied along the girth and length of the container in order to provide additional strength to the container. Some of the most commonly used are wooden pallets, metal straps and plastic straps.

Requirements of a Consumer Pack

- ☆ Adequate protection
- ☆ Details about the product like materials used for the production of the product(ingredients in the case of food products)
- ☆ Details about the content.
- ☆ Mandatory requirements like date of manufacture, address of manufacturer, date of expiry, or date well before use, brand name, graphics, measure of the content, license number, authority issuing the license.
- ☆ Provision for inspection before purchase.
- ☆ Easiness in opening
- ☆ Re closing facility
- ☆ Easiness in grip
- ☆ Measurable dose if possible
- ☆ Easiness in disposability
- ☆ Attractive to promote sales
- ☆ Cost effective

Classification Based on Packaging Techniques

Vacuum Packaging

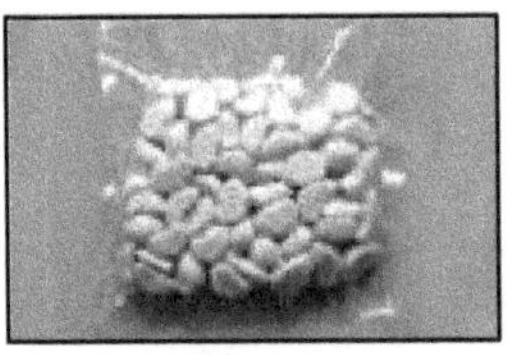

Vacuum Packed Nuts

This is the simplest and most common method of modifying the internal gaseous atmosphere in a pack, the product is placed in a pack made of plastic film or laminate of low oxygen permeability, air is then evacuated and the package is sealed. Evacuated package collapse around the product. The main significance behind the technique is to extend the shelf life of the processed food products. Products having high fat content like fried items become rancid due to reaction with oxygen. As a result of rancidity the product will become bitter and un acceptable to the consumer. In order to reduce the oxygen available within the package it is needed to evacuate the air from inside the package. This technique is most suitable for powdery products like skimmed milk powder, tea dust etc and not so suitable for granular or flaked or bakery products as the products get crushed due to evacuation of air from the package.

Gas Packaging

Nitrogen Filled Gas Pack for Fragile Materials like Potatochips

Gas packaging can be achieved either mechanically by evacuating the air and then replacing the space with a gas or mixture of gases, or other way by generating the atmosphere either passively as in the case of fruits or vegetables or actively by using suitable atmosphere modifiers such as oxygen absorbents, carbon di-oxide absorbents or emitters. In general gas flushing process is usually done on a vacuum packaging machine and the gas in various proportions is injected into the package by the same machine attached with the gas cylinders. This dilutes the air in the head space surrounding the food product. Inert gas like nitrogen can be used to fill the package then it is called nitrogen packaging or if it is carbon di oxide it is called carbon di oxide package. Apart from the chemical or microbiological point of view in protecting the product, it will also provide physical property like cushioning or avoidance of direct pressing to those fragile products like potato chips from mechanical damage. In certain cases carbon di oxide package will prevent the growth and development of store pests in dried products like cashew nuts.

Modified Atmosphere Packaging (MAP)

Modified Atmosphere package modifies the internal atmosphere of food. It is primarily applied to fresh or minimally processed foods that are still undergoing respiration. It is also used as a packaging technique for baked goods, coffee beans, tea, dairy products, dehydrated foods, processed meat to keep the meat pigment looking desirable, it is also used for nuts and snack food application.

Minimally Processed Vegetables in MAP

MAP contains the food under a gaseous environment that differs from air in order to control normal product respiration and growth of aerobic micro organisms. Nitrogen gas which is odourless, tasteless colourless, non toxic and non inflammable is most widely used for modification of the atmosphere. This modification offers protection from spoilage, oxidation, dehydration, weight loss and freezer burn and extends the shelf life.

Controlled Atmosphere Packaging (CAP)

Both controlled atmosphere in storage (CAS) or in package (CAP) permit only controlled oxygen and carbon dioxide exchange. The gases mainly controlled in CAP or in CAS are oxygen, carbon dioxide, ethylene concentration, water vapour etc for high quality food for world wide distribution.

Active Packaging

Active packaging is a group of technologies in which the package is involved with the food products or interacts

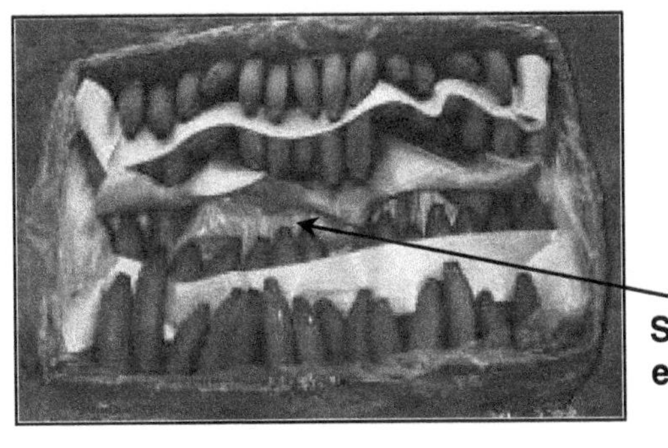

Banana Hands in an Active Package

with the internal atmosphere to extend the shelf life, provides physical barrier to external spoilage, contamination and physical abuse in storage and distribution. Active package contributes to product development, controls maturation and ripening, helps in achieving the proper colour development in meats.

In the case of packaging fresh and minimally processed foods it provides moisture or oxygen barrier to control loss of moisture and enzymatic oxidative browning in fresh cut fruits and vegetables and to provide controlled permeability rates matched to respiration rate of the fruit.

Examples for Active Packaging

- ☆ Films that are scavengers of off odours.
- ☆ Oxygen scavengers for low oxygen packaging.
- ☆ Fumigation in case of grapes like grape guard against wild yeast.
- ☆ Microwave suspected films to allow browning and crispness (French fries, baked products, popcorn) etc.
- ☆ Steam release films for self heating.
- ☆ Time-temperature indicators which are unable to reverse their colour when the product has been subject to time- temperature abuse for frozen products.
- ☆ Ethylene scavenging films
- ☆ Microbial growth indicators.
- ☆ Light protection films (photochromic)
- ☆ Physical shock indicators
- ☆ Leakage indicators.

Active packaging can be called as "smart packaging" also, the package function switches on and off in response to changing external/ internal conditions and can include a communication to the consumer or end user as to the status of the product. A simple definition of intelligent packaging is "packaging which senses and informs" various aspects like safety and disposal instructions *e.g.* Cox Technologies has developed a colour indicating tag that is attached as a small adhesive label to the outside of packaging, which monitors the freshness of sea food products. A barb on the back side of the tag penetrates the packaging film and allows the passage of volatile amines, generated by spoilage of the sea food which will progressively turn the fresh tag colour into pink as the sea food ages.

Self heating packages for soup and coffee, self cooling containers for beer and soft drink are some of the other active packages. Intelligent Packaging can be defined as: Packaging systems that are capable of carrying out intelligent functions (such as detecting, sensing, recording, tracing, communicating, and applying scientific logic) to facilitate decision making to extend shelf life, enhance safety, improve quality, provide information, and warn about possible problems.

In short, it is "a'packaging which senses and informs".

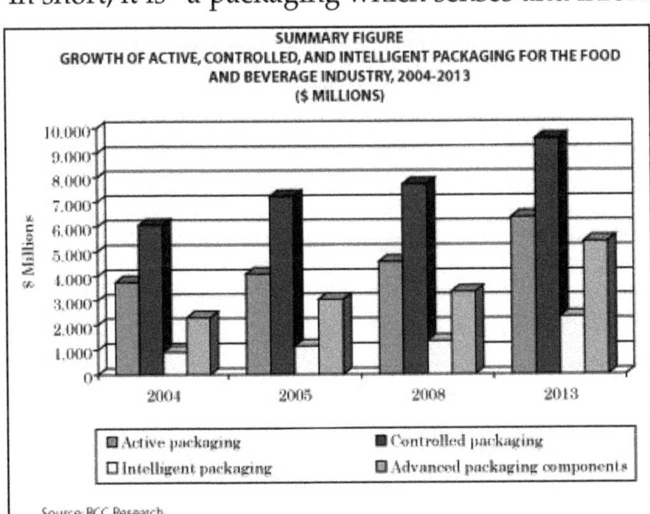

Why Intelligent-Packaging?

- ☆ Retain integrity and actively prevent food spoilage (shelf-life)

- ☆ Enhance product attributes (look, taste, flavour, aroma etc.,)

- ☆ Respond actively to changes in product or package

- ☆ Communicate product information, product history or condition to user

- ☆ Assist with opening and indicate seal integrity

- ☆ Confirm product authenticity and act to counter theft.

What an Intelligent Package Does?

- ☆ Sense the environment and convey appropriate information to the user about the contents of the package

- ☆ Give information about the quality of the packaged food during transport and storage

- ☆ Facilitate decision making to achieve the benefits of enhanced food quality and safety

- ☆ Sense the environment and convey appropriate information to the user about the contents of the package

- ☆ Give information about the quality of the packaged food during transport and storage

- ☆ Facilitate decision making to achieve the benefits of enhanced food quality and safety.

Internal and external sensors, which are connected to Radio-Frequency Identification (RFID) tags gives information about the state of the product as follows:

New thermal barcodes-track quality of food in cold chain. Using nano-technology indicators help removing expired items from store shelves. Computer chip/label attached with the package to:

- ☆ Identify raw materials in the package
- ☆ Determine whether it has been stored at proper temperature for the right amount of time.
- ☆ Tell us about pricing, labeling and packaging information.
- ☆ When indicators are used in milk cartons it will:
    1. Determine freshness level
    2. Shelf-life limits
    3. Vitamin D content
- ☆ When in Fresh foods
    1. It will detect in the production of gas, moisture and micro organisms
- ☆ In Bread and pizza it will detect the mould growth
- ☆ In vegetable oils it will detect the rancidity and loss of flavour
- ☆ Intelligent packaging shows whether they are spoiled due to temperature change during storage or a leakage in packaging.

"Ripe Sense" Labels

The "Ripe Sense" sensor labels that enables shoppers to choose fruit, that best appeals to their taste. It works by

Fruit Packed with "Ripesense" Label
Red: Unripe; Orange-red: Half-ripe; Yellow: Fully ripe

detecting aroma compounds given off the fruit as it ripens, changing the label through a range of colours.

Fish

The spoilage of fish is indicated by the production of Try Methylnmine (TMA). TMA is produced only at the later stages of the spoilage causing bad smell, so difficult to use as indicator. In fresh fish where as ammonia and ATP degradation products like aldehydes, ketones etc. are formed in the early stages of the storage which is used as a sensing agent to detect the spoilage in fish

Meat and Fish

There is a chemical in the package label that changes colour from light to dark when it reacts with the volatiles produced as a result of spoilage. If the packet is kept cool, the reaction is slow, but increased temperature speeds up the reaction. As bacteria on food also grow faster if it is warm, it is useful to know if food has been kept cool or not. The 'intelligent' label alerts consumers if the meat on sale has been kept too warm for too long. Once the label has gone dark, the product is not guaranteed to be fresh.

Another new type of label is being developed for use on foods that give off gases when they decay. The labels will change colour if they detect the gases so we can tell if the food is fresh before we open the pack. Other developments include food wraps that kill bacteria and a sandwich wrap which changes colour if it detects the presence of harmful bacteria.

Intelligent labelling and printing communicate directly to the customer via thin film devices providing sound and visual information, either in response to touch, motion or some other means of scanning or activation.

Voice-activated safety and disposal instructions contained on household and pharmaceutical products–tell the consumer how they should be disposed off after consumption–information directly used in the recycling

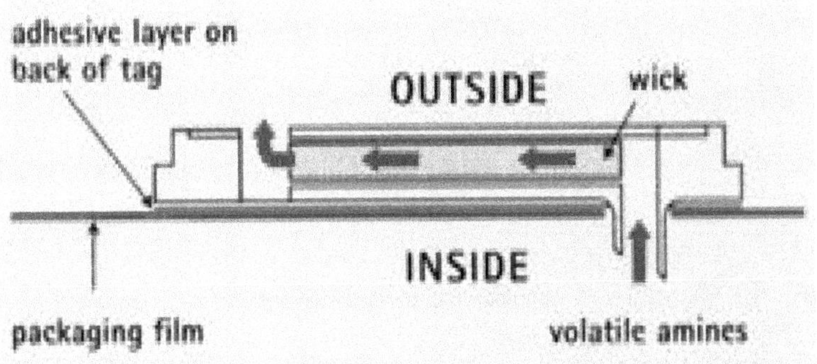

industry to help in sorting packaging materials from waste streams.

Drug delivery systems in smart packaging will be programmed to communicate patient's information back to healthcare centres.

COX Technologies has developed a colour indicating tag that is attached as a small adhesive label to the outside of packaging film, which monitors the freshness of seafood products.

A barb on the backside of the tag penetrates the packaging film and allows the passage of volatile amines, that turns 'Fresh Tag' progressively bright pink as the seafood ages.

Fresh Fruits and Vegetables

Novel breathable polymer films are in commercial use for fresh-cut vegetables and fruits. Acrylic side-chain crystallisable polymers tailored to change phase reversibly at various temperatures from 0-68°C. The final package is 'smart' because it automatically regulates oxygen ingress and carbon dioxide egress by transpiration according to the prevailing temperature. In this way, an optimum atmosphere is maintained around the product during storage and distribution.

Tamper-proof Packaging

Tamper evidence technologies, based on optically variable films or gas sensing dyes, involve irreversible colour changes, cost-effective for disposable packaging of commodity items. Piezoelectric polymeric materials might be incorporated into package construction so that the

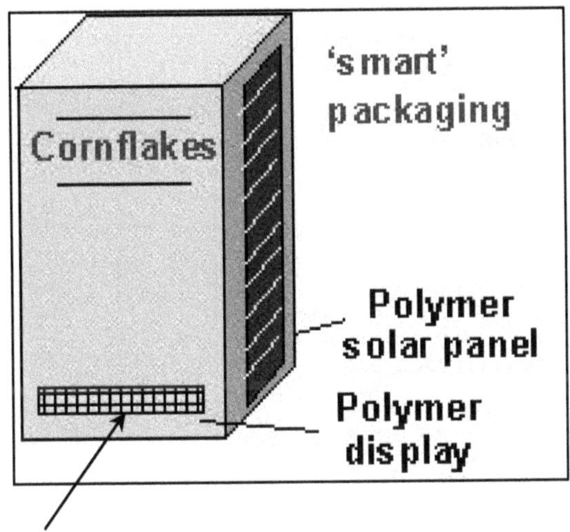

Polymeric Material which will Change Colour at Certain Stress Threshold Levels indicating the Spoilage of the Product

package changes colour at a certain stress threshold. In this way, a 'self-bruising' closure on a bottle or jar might indicate that attempts had been made to open it.

Why Intelligent Packaging has a Bright Future?

☆ Assurance of freshness and safety of food will increase consumer's demand

☆ Globalization and expansion of marketing make logistic chain longer, placing more demands on traceability

☆ Facilitation of in-home control for industry and retailing

☆ Monitor product quality and trace critical points in food supply chain

Aseptic Packaging

Aseptic packaging has been defined as a procedure consisting of sterilization of the packaging material or container, filling of commercially sterile product in a sterile environment and producing containers which are tight enough to prevent contamination. The term "Aseptic" implies the absence or exclusion of any unwanted organisms from the

Aseptically Packed Fruit Juice in Tetra Brick Pack

product, package or other specific areas while the term "hermetic" is used to indicate suitable mechanical properties to exclude the entrance of bacteria into the package or more stringent to prevent the passage of micro organisms, gas or vapour into or from the container.

Most commonly used packaging material for aseptic packaging is the Tetra pak, which is a laminate comprising layers of paper, polythene and aluminum foil. Paper makes the package stiff, plastic makes the water tightness and aluminum foil protect the contents against light and oxygen. "Tetra Pak" has developed a packaging system that can preserve products without the use of refrigeration. Fluid products can be preserved in this process, the product is heated for a very brief period at an ultra high temperature (UHT), then flows through a closed system to the packaging machine, where it is filled into a sterile packaging material under completely sterile conditions. Packaged in this manner, the product retains its taste and nutritional value.

Edible Packaging

An edible packaging is a thin layer that is deposited on the surface of a food or in between food components and is co-consumed.

Edible Chocolate Tray made from Corn Starch

Edible Films

Uses it as:

☆ Film to wrap foods

☆ Spray or dip to coat foods

Main Classification of Edible Films

1. Biopolymers
2. Edible lipids
3. Composite films

1. Biopolymers (Proteins and Polysaccharides based)

Hydrophilic in nature. Therefore good barriers against hydrophobic compounds like lipids, oxygen and certain flavours.

e.g. Gluten, milk protein, gelatin, starch, pectin, cellulose-ethers, corn zein etc.

Edible Protein

Whey Protein (A common material used for coating in the edible films are due to the following reasons):

☆ A by-product of cheese manufacture and hence a cheap raw material.

☆ Used to make edible film and coating that are oxygen, aroma, oil and moisture barriers.

☆ When coated has got good mechanical properties.

- ☆ Durability when used as coating on food products.
- ☆ Films will be separating the layers of heterogeneous foods.
- ☆ Pouches can be made for food ingredients.
- ☆ Recyclability of the packaging waste can be improved.

Edible Lipids

- ☆ Hydrophobic in nature.
- ☆ Good barrier against water and water soluble compounds

  e.g: Waxes,triglycerides, shellac, fatty acids, sucrose, fatty acid esters; acetylated monoglycerides are the most commonly used edible lipids.

Composite Films

Consisting of a blend of polysaccharides, proteins or lipids. It improve the mechanical property.

e.g. Lipid, hydroxypropyl methyl cellulose, corn zein (protein), methyl cellulose, fatty acids.

Importance of Composite Films

If we want barriers against both hydrophilic as well as hydrophobic compounds it can be used as composite films. For example barriers against water and oxygen we have to use combinations of lipids and biopolymers as:

1. Bilayer coatings.
2. Emulsion based coating.

Possible Uses of Edible Film and Coatings

- ☆ Improve structural integrity and handling properties.

☆ Retain volatile flavour components.

☆ Convey food additives.

Purpose of using edible coatings are:

☆ Retard oil and fat migration.

☆ Retard moisture migration.

☆ Retard solute migration.

Characteristics

☆ Prevent moisture migration

☆ Extend the shelf life-by acting as barrier to moisture, oil, vapour transmission and aroma.

Advantages

☆ Provide protection to the inner contents.

☆ Improved handling properties like melt in our mouth but not in our hand.

☆ Enrich the film with extra nutrients such as vitamin-E, calcium etc., it boost the nutritional value of the food.

☆ Reduce wastage means; cutting down the amount of packaging required such as bag-in-box, plastic-plus-foil laminates.

☆ Impart flavour to products such as seasoning to crude meat.

☆ They adhere to the food and not the packaging material.

☆ Good eating properties like acceptable colour, taste, flavour and texture.

☆ Decrease the conventional synthetic packaging materials

☆ Enhance the appearance

Disadvantages

☆ Production cost is high

☆ Not good barriers against insects and micro-organisms

☆ Chemical composition of the food and the film must match otherwise it may for produce adverse effect

☆ Bee wax tends to crack, or form a coating so thick, that it disrupts respiration in the cells thus causing "off" flavour development due to an aerobic respiration.

Plasticizers

☆ Edible film often requires incorporation of low molecular weight plasticizers to improve film flexibility and durability.

☆ By interrupting the chain interactions or by lowering the glass transition temperature

e.g. Edible plasticizers include Sucrose, Glycerol, Propylene glycol, Fatty acids, monoglycerides

Method of Producing an Edible Film

Technologies used for producing edible films are similar to that of thermoplastic structures. Conditions are different but the principles are allied. Solvent casting and extrusion for water soluble edible materials. Water or water-ethanol solutions or dispersions of the edible materials are spread on smooth surfaces. When the solvent is evaporated,the film can be stripped off from the surface. Water insoluble collagen films for sausage casings is made by regeneration of collagen. Formation of collagen film involves extrusion of a low-solids content collagen in to a coagulation bath. By products from dairy processing industry as well from bio-fuel is used to create a biodegradable protective films. A film made from milk protein is having an excellent water resistant property. Another edible film made from starch is widely used in food and medicine industries whose advantages are its resistance to acids, alkali, heat and moisture. It is also having an excellent transparency.

Biopolymers are those obtained by the microbial action on low cost nutrients. *e.g.* Dextran.

Dextran is obtained by the fermentation with *Leuconostoc mesenteroides* isolated from a typical Mexican beverage named "pulque" (cactus juice fermented). Dextrans were purified by precipitation with methanol. The functional properties of Dextrans were compared with four highly purified biopolymers. The best efficiency of Dextrans resulted when coated fruits were preserved at 4°C, by increasing the quality characteristics like size, colour, aroma and water content.

Use of Biopolymers in Meat

Hydroxypropyl methylcellulose an edible film coating on chicken meat balls for moisture retention and fat reduction during deep-fat frying is an example for the typical commercial application of biopolymers.

Cakes Decorated with Edible Substances

The coating is made of chitosan and lysozyme. Chitosan is a fibre found in crab and shrimp shells and lysozyme is a protein found in egg whites. The concept is that if the coating is edible then nutrients can be placed in the food coating that can then act as a way to fortify the food.

Functional Properties of the Coatings

1. Control mass transfer (Preventing dehydration): Venting food from desiccation, regulating micro-environment of gases around food, controlling migration of ingredients and additives in the food system.

2. Provide adequate mechanical strength

3. Protect the integrity of packing throughout distribution: Edible packaging refers to the use of edible films, coatings, pouches and bags and other containers as a means of ensuring the safe delivery of food product to the consumer in a sound condition.

Generally an edible film is defined as a thin layer of edible material formed on a food as a coating or placed between or on the food components. Its purpose is to inhibit migration of moisture, oxygen, carbon-dioxide, aroma and lipids.

Effective edible films formed as coatings on foods by dipping, spraying or painting could reduce packaging requirements and waste. An edible film coating act as an efficient moisture, oxygen or aroma barrier materials, which can reduce the amount of packaging material. Various functions of edible coatings are:

Barrier Property

One of the most useful functions of edible film is their ability to act as barriers, either to gas, oil or more often water. Moisture levels in foods are critical for maintaining freshness, controlling microbial growth and providing mouth feel and texture. Edible films can control water activity preventing either moisture loss or uptake.

Water dispersible forms of corn protein can be applied as a film coating to provide a moisture or gas barrier for nut or fruits. One such application would be coating raisins for use in dry ready- to- eat breakfast cereals. Here the zein coating prevents moisture migration from the raisins to dry cereal helping to maintain their quality.

Besides being moisture and gas barriers, edible coatings can act as barriers to oil uptake. An example for the use of pectin as an oil barrier in French fries, since pectin binds potato cells together it would improve resistance to oil penetration. Similarly the use of plant gum based coatings have been used for several years in Japan and other Asian countries again for preventing oil absorption in French fries.

Binding

Second function of using edible coatings to surface of snack foods and crackers to serve as a foundation or adhesive for seasonings. Application of modified food starches in combination with corn syrup, water and glycerine are used to make an adhesive solution, this solution is used as a coating for peanuts then seasonings or salt may be added to the surface as an edible coating.

Glaze

Third function of edible film is to act as glazes to enhance the appearance of baked goods. Wheat gluten film avoids the possible microbial problems associated with raw egg products and provides some barrier properties against moisture loss.

Safety and Health Issues

Determination of the acceptability of materials for edible polymer films follows procedures identical to determining the appropriateness of such materials for food formulations. As edible polymer will be Generally Recognized as Safe (GRAS) to use the edible films if the material has previously been determined as GRAS and its use in edible film is in accordance with current good manufacturing practices (food grade, prepared and handled as food ingredient and used in amounts not greater than necessary to perform its function) and within any limitations specified by the Food and Drug Administration (FDA).

The materials that have received the greatest attention for edible film use are cellulose ethers, starch, corn zein, wheat gluten, soy protein and milk proteins. Attention to the microbial safety of edible film is guided by standard considerations of water activity, pH, temperature, oxygen supply and time. Importantly edible films are effective carriers of antimicrobials which improve the microbial stability of film and food alike.

Biodegradable Polymeric Films

Biodegradable plastics are the materials that are in combinations of natural and synthetic material *e.g.* starch with LDPE. Although these materials disintegrate in composting operations, they did not completely biodegrade. In the real sense a material to be called biodegradable they must be degraded completely by micro organisms in a composting process to natural compounds like carbon-dioxide, water, salts etc. involving three key elements-appropriate micro-organisms, environment with acceptable micro-organisms and conducive to efficient bio degradation process.

Replacing conventional synthetic packaging with bio degradable polymers can reduce the use of non renewable resources and decrease waste through biological recycling in to a bio system.

FDA regulates all materials proposed for food packaging to ensure they are safe for food contact under conditions of intended use. Normally the bio degradable packaging material developed must be under GRAS (Generally Recognized As Safe). FDA requires the information of the bio degradable material on the migration profile and identity of migrants from bio degradable packaging to food during typical storage conditions and times.

Non-edible polymeric materials are made available as biodegradable polymer films by including certain cellulose based products *e.g.* Cellophane, microbial polyester *e.g.* polyhydroxy butyrate, biodegradable synthetic polymers *e.g.* polylactic acid and combinations of starch with, bio degradable synthetic polymers *e.g.* polyvinyl alcohol.

The challenge for the successful use of biodegradable polymer products is achieving controlled life time, means the products must remain stable and function properly during storage and intended use and biodegrade efficiently later.

Biodegradable but not edible cellophane makes a strong package due to its good tensile strength and elongation and tears easily when notched. Other attributes include excellent printability and good machinability. Like other natural polymers cellophane is sensitive to moisture because of these inherent hydrophilic nature, it is not heat sealable. So cellophane is often coated with nitrocellulose wax or with polyvinylidene chloride (PVDC) and is often additionally laminated *e.g.* metallised polyester,when cellophane is coated with nitrocellulose it is totally bio degradable but coating with PVDC it is not fully bio degradable.

Cellulose acetate a thermoplastic material is crystal clear and tough and is excellent for certain high moisture products because it breathes and does not fog up. It is also a good barrier to grease and oils. Although chemical substitution of cellulose generally slows biodegradation, cellulose acetate appears to be biodegradable.

The usefulness of cellulose as a starchy material for edible and biodegradable films can be extended by chemical modification to methyl cellulose(MC), hydroxyl propyl methyl cellulose (HPMC),hydroxyl propyl cellulose(HPC) and carboxy methylcellulose(CMC).These cellulose ether films have moderate strength, resistance to oils and fats. They are flexible, transparent, odourless, tasteless, water soluble, moderate barrier to moisture and oxygen.

Polysaccharide, Starch Based Biodegradable Materials

Starches can be used to form edible and bio degradable film. Amylose, high amylose starch and hydroxyl propylated high amylose starch have been used to form self supporting films by casting from aqueous solution. These films appear to have moderate oxygen barrier properties but are poor moisture barriers and their mechanical properties are generally inferior to synthetic polymer films. The influence of moisture, however on the stability of starch film limits their usefulness.

Starch Composites

Starch-polyethylene mixtures can be processed via

extrusion, injection molding, film blowing for the production of films or bottles. The starch component of these films can be metabolized by certain amylolytic bacteria, but the polyethylene component does not biodegrade, rather it disintegrate.

Starch and polyvinyl alcohol (PVOH), a biodegradable synthetic polymer offering a broad range of properties have been blended to yield thermoplastic materials with properties superior to starch alone, some starch-PVOH blends appear to have potential for replacing LDPE films in applications where mechanical properties are critical for the intended use and good moisture barrier properties are not necessary.

Other Polysaccharide Materials

Alginate films are formed by evaporation of an aqueous alginate solutions followed by ionic cross linking with a calcium salt. They are impervious to oils and fat but are poor moisture barriers. Despite this, alginate gel coatings can significantly reduce moisture loss from foods by acting sacrificially. In other words, moisture is lost from the coating before the food significantly dehydrates. Alginate coatings are good oxygen barriers, can retard lipid oxidation in foods and improve flavour, texture and better adhesion.

Coatings that include carrageenam as a major or sole component have been applied to a variety of foods to carry anti microbial and to reduce moisture loss, oxidation or disintegration. Protein based collagen is a fibrous, structural protein in animal tissue that can be converted into edible and bio degradable films. Because collagen is not thermoplastic, collagen must be made by extruding a viscous colloidal acidic dispersion into a neutralizing bath followed by washing and drying. Collagen film is not tough as cellophane but has reasonably good mechanical property. It has an excellent oxygen barrier at 0 per cent RH, but oxygen permeability increases rapidly with increasing RH in a manner similar to cellophane. Collagen film over wrap on refrigerated and thawed beef round steak, packaged in high barrier bags or in PS foam trays with polyvinyl chloride (PVC) film reduced exudation without significantly affecting the colour or lipid oxidation another study found that collagen film performed as good as plastic film in maintaining the quality of frozen beef cubes. Collagen film, unlike synthetic polymeric film melts away as the meat thaws and cooks, eliminating the need for plastic film waste handling.

Wheat gluten has been exploded as a plant based replacement for collagen in the manufacture of sausage casings as a means to improve the adherence of salt and flavourings to nuts and batters for meat and other foods. Soy protein has also been studied for the manufacture of sausage casings and in the production of water soluble pouches. Soy protein in edible coatings can improve better adhesion and reduce moisture migration in raisins and dried peas.

Whey protein coatings effectively carried antioxidants for frozen fish, significantly reduced oxygen uptake and rancidity in roasted peanuts and reduced disintegration of freeze dried foods.

The hydrophilic nature of edible polymers limits their ability to provide desired edible film functions. For all edible polymers, RH greatly influences the properties. Use of edible polymer films and coatings as moisture barriers usually requires the formation of composite films that contain hydrophobic materials such as edible fatty acids and waxes. Edible polymers that are water soluble and emulsifiers may be favoured in many food coating applications. Development of composite bi-layer films will allow protection of edible film oxygen, aroma and lipid barrier properties with a hydrophobic moisture barrier layer.

When considering possible edible film coating applications, attention must be given to the requirement that edible coating formulations must wet and spread on the food surface and upon drying form a film coating that has adequate adhesion, cohesion and durability to function properly. These properties are influenced by both edible film formulations and method of coating and drying. In addition, edible coating must provide satisfactory appearance, aroma, flavour and mouth feel. Selective application of appropriate foods and good control of environmental conditions are necessary to ensure microbial stability.

Bio-plastics

Bio-plastics are another sort of bio degradable plastics. Bio-plastics are defined as polyesters produced by a range of microbes, cultured under different nutrient and environmental conditions.

Why Bio-plastics

Availability of raw materials from agriculture and marine feed stocks can be utilized as source for growing micro organisms.

☆ Bio-plastics are again biodegradable.

☆ Preserve non renewable resources–petroleum, natural gas and coal.

☆ Waste disposal management is easy

Degradation Mechanisms

☆ Biodegradable

☆ Compostable

☆ Hydro-biodegradable

☆ Photo-biodegradable

☆ Bio-erodable

Biodegradable Mechanism

As defined by American Society of Testing Materials (ASTM) Biodegradable plastics are those "Capable of undergoing decomposition into carbon-dioxide, methane, water, inorganic compounds or biomass in which a predominant mechanism is the enzymatic action of micro-organism that can be measured by standardized tests, in a specified period of time, reflecting available disposal condition."

Compostable plastics are those "Capable of undergoing biological decomposition in a compost site as apart of an available programme, such that the plastic is not visually distinguishable and breaks down to carbon-dioxide, water, inorganic compound and biomass, at a rate consistent with known compostable material (*e.g.* cellulose)".

Hydro-degradable

☆ Polymers are broken down in two step processes.

☆ Initial hydrolysis and further biodegradation.

Photo-biodegradation

☆ Polymers are broken down in two step process

☆ Initial photo-degradation followed by biodegradation.

Bio-erodible

Abiotic Disintegration

☆ Initial degradation is without action of micro organism.

☆ Pretreatments include dissolution in water, heat ageing, UV ageing.

☆ Standards for biodegradability

Each country has its own standards.

Requirement–90 to 60 per cent decomposition of the product within 60 to 180 days of being placed in a standard composting environment is supposed to be a standard de gradation process.

Environmentally Degradable Plastics

Broken down into simpler substances by living organisms or natural sources. Natural plant polymers derived from wheat or corn starch have molecules that are readily attacked and broken down by microbes.

Biodegradable Starch Based Polymers

e.g. Thermoplastic starch products.

Starch synthetic aliphatic polyester blends.

Starch-Poly vinyal hydreside.

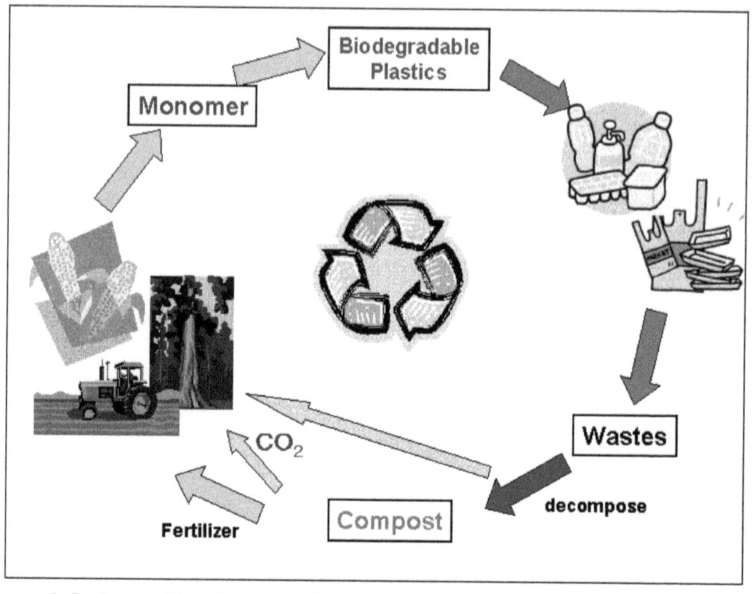

**A Schematic Diagram Illustrating the Decomposition of
a Biodegradable Plastic**

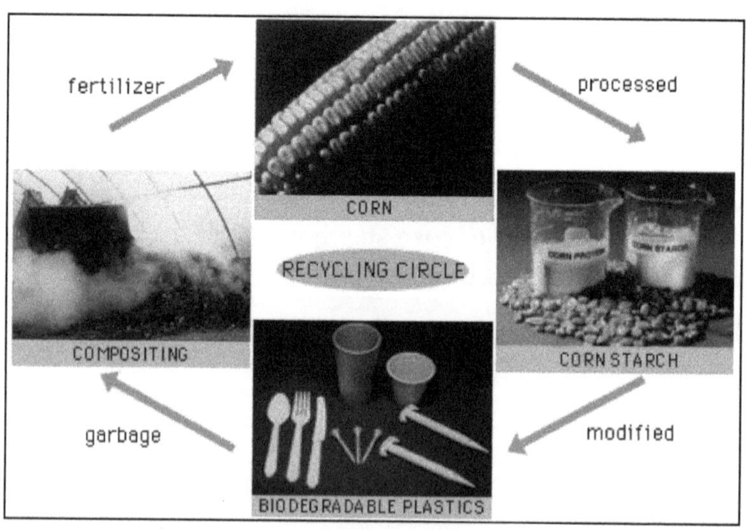

Recycling of Biodegradable Plastic

Plastics from Bacteria

e.g. Polyhydroxyalkanoate (PHA) a granular plastic

Plastic from other sources:

Gelatin–collagen (Used as drug capsule)

Casein–natural protein (Used as binders and protective coatings) for cellulose.

Range of Degradable Plastics

Starch based products

Naturally produced polyesters

Synthetic aliphatic polyesters

Aliphatic – aromatic (AAC) co-polymers

Modified PET

Water soluble polymers

Photo-degradable

Commercial Development

Polyhydroxybutyrate (PHB)

Polylactic acid (PLA)

Polycaprolactone (PCL)

Polybutylene succinate (PBS)

Polyhydroxyhexaoate (PHH)

Polyhydroxyvalerate (PHV)

Application and Uses

Packaging–shopping bags, trays, containers, bottles, blister foils.

A Carry Bag of Biodegradable Material

Other Applications of Bioplastics

Medicated syringes; Composting bags

Capsule coating; Food service cups

Sanitary products; Disposable razors.

Shopping bags

Suture threads

Crocery bags

Food packaging.

Injection moulding

Biodegradable plastics will replace traditional plastics soon with up coming advancement in this field.

Antimicrobial Packaging Materials

Addition of antimicrobials to edible films may have wider applications and extends the shelf life. When antimicrobial agents are incorporated into a polymer in the form of films, containers or utensils the material will then limit or prevent the microbial growth.

A chemical preservative can be incorporated into a packaging material to add antimicrobial activity. For example preservative releasing films provide antimicrobial activity by controlled oxygen rate. Oxygen absorbents also reduce head space and thus partially protect food against aerobic spoilage such as mold. Common antimicrobial chemicals added to packaging materials or edible coatings are organic acids and their salts, sulphites, nitrites, antibiotics and alcohol.Eg. sorbic acid and potassium salts have been used along with the packaging of cheese products. They were mixed into a wax layer for natural cheese and coated on to the packaging material. LDPE films into which benzoic an hydroxide incorporated exhibited anti mycotic activity. Addition of 0.5 to 2 per cent benzoic anhydride to LDPE films has delayed the mould growth in cheese.

Nisin as an antibacterial compound has been accepted for antimicrobial application in packaging and has been found effective against *Clostridium botulinum*. Incorporating radiation emitting films into food packaging material by Japanese have reportedly developed materials that emit long wave infrared radiation (IR) on exposure to water or water vapour and found effective against microorganisms. Antimicrobial enzymes such as glucose oxidase when coated to the inner surface of the contact-films will produce antimicrobial property.

Metallic ions like silver ion has the strongest antimicrobial activity compared to other metal ions like sodium, potassium, iron, zinc, nickel etc. Silver is a safe and relatively an inert material and therefore often used in direct human contact with dishes, plates, forks, spoons, knives and tooth fillings.

Vegetable parchment paper is coated with sorbic acid and is used as a base paper for the packaging of bread and has found to have a mycostatic effect in foods. Generation of chlorine di oxide within the package interior from solid state technology is activated by one or more of the ambient environmental conditions such as humidity. The duration and amount of chlorine dioxide generated can be modulated to kill broad range of micro organisms including mold spores. These polymer films were successfully used to destroy *Escherichia coli* in ground fresh meat.

Nylon film was treated with ultra violet irradiation to increase surface amine concentration which act as surface active antimicrobial sites. Ethanol sachets containing encapsulated ethanol will release ethanol vapour, which impart a preservative effect in the packaging head space. Ethanol prevents microbial spoilage of intermediate moisture foods like cheese and sweet bakery goods.

The antimicrobial activity of incorporated active substance can be deteriorated during:

Casting: *i.e.* fabricating into containers

Lamination: *i.e.* bonding with other materials

Printing: *i.e.* providing graphics or addresses

During storage and distribution–due to exposure to adverse temperature and humidity.

Therefore extra precaution must be taken while material coated with antimicrobial aspects to retain its residual microbial activity, when the food is packed to serve its purpose

References

Abhijit Kar (2005) Disposal of packing materials in the lecture schedule on Food Processing and Engineering, IGNOU, New Delhi.

Basantias (2000) Milk proteins in the preparation of edible coatings, Indian Food Industry, 19910: 36-45.

Butler P. (2001) Possible concerns over intelligent packaging, Materials World Vol. 9(3).

Khamrui,K. (1999) Edible packaging: An ecofriendly alternative to the plastics, Indian Food Industry 18(1) 34-38.

Saha,N.C.(2005) Biodegradable plastics in the lecture Schedule for Food Processing and Engineering, IGNOU New Delhi.

TanweerAlam (2005) Antimicrobial packaging, Processed Food Industry, (2)" 20-24.

http://www.wipo.int/petdb/en/wo.jsp.

http://www.dairy science.info/map-science.asp.

http://www.ediblefilms.org/.

http://www.nytimes.com/2007/08/29/dining/29film.html.

http://www.freepatentsonline.com/EP1586243.html.

http://www.freshplaza.com/news_detail.asp?id=34964.

http://www.alibaba.com/catalog/Edible_Film_Strips.html.

http://www3.interscience.wiley.com/journal.

http://www.organicconsumers.org/foodsafety/packaging122004.cfm.

http://www.actahort.org/members/showpdf.booknrarn.

http://www.sciencedaily.com/releases/2007/06/07062.

Chapter 4
Terminologies Used in Packaging

1. Thickness

Thickness of a material is the perpendicular distance between the two outer surface of the material and is normally expressed in units of length. Many physical properties of packaging materials are dependent upon the thickness *e.g.* Water Vapour Transmission Rate (WVTR) and Gas Transmission Rate(GTR) of a film is inversely proportional to thickness and decrease with increase in thickness.

Dial gauge, micrometer, screw gauge, vernier calipers etc are used for the measurement of thickness.

For paper boards, thickness is reported in points or in mm (1 point = 1/1000 of an inch); for papers it is in mm or inches.

For films, thickness is reported in micron, mils or in gauge (25 micron = 1mil = 1/1000 of an inch = 100 gauge = 0.25 mm).

2. The Basis Weight for Paper or Paper Boards

The basis weight is the average weight of an arbitrarily selected area of the paper (weight per unit area in grams/ sq. meter or lbs/1000 sq ft). As the packaging papers are sold and purchased only in terms of weight, the basis weight assumes special significance. Most of the physical properties such as burst strength, thickness and bulk are evaluated and specified in accordance with the particular basis weight involved.

3. Tensile Strength

The tensile strength of a paper is defined as the force applied parallel to the plane of the specimen of specified width and length under specified condition of loading (kg/ 15mm or lbs/inch width). The test indicates the durability and serviceability of papers in many packaging operations such as wrapping, bagging, printing etc. usually tensile strength is more in machine direction than in transverse direction and extension is less in machine direction than in transverse direction.

Plastic films are normally tested at higher speeds because of higher extensibility. The stress strain curve helps in locating the yield point and knowing the yield strength

4. The Burst Strength

The test measures the ability of a paper or paper board to withstand pneumatic or hydraulic pressure built up. For films, foils, laminates and papers the pneumatic test is used.

Heavy papers and paper boards are tested on hydraulic type of testers (lbs/sq. inch or kg/ sq. cm)

The test gives a sort of combined tear and tensile properties. In many cases it serves as good index of the quality of fabrication of packaging materials.

5. Water Vapour Transmission Rate (WVTR)

The WVTR is measured as the quality of water vapour in grams that will permeate from one side to other side of the film of an area of one square meter in 24 hours, when the relative humidity difference between the two side is maintained at 90 per cent gradient at 37.8°C. The property is important to estimate the efficiency of the packaging material or a package for resistance to the flow of water vapour and is helpful in considering the selection of barrier materials for hygroscopic foods.

6. Gas Transmission Rate (GTR)

The GTR is normally determined by measuring the change in pressure at constant volume and the quantity of gas flowing across the film is compiled as volume at NTP.(Normal Temperature and pressure)

$$GTR = \frac{V\,(76) \times 24}{At\,(p1-p2)} \quad \text{As CC/sqm/24 hrs at atmospheric pressure}$$

Where 'V" is the volume (at NTP) of gas transmission through "A" unit square meter of the test material in time "t", when the average pressure difference between the two sides is maintained at (p1-p2) cm of mercury. The temperature of the test can be changed as per the requirement. GTR is an important property to estimate the efficiency of the packaging material or a package resistance to the flow of gases and helps in selection of barrier materials for oxygen sensitive foods.

7. Grease Resistance

Grease resistance is measured by exposing one of the test specimen creased or un creased to a grease containing red dye. The time required for the red stain to show on the unexposed side is taken as a measure of this property. For plastic films, the test can be performed directly in pouches using ground nut oil coloured with red Sudan dye.

8. Tear Resistance for Papers

The papers are tested for their tearing resistance properties in two ways:

Internal tearing: The energy required to propagate an internal tear is measured.

Edge tearing: The energy required to initiate a tear is measured.

The test is done on both directions of the paper. The work done in tearing is measured by the loss in potential energy of the pendulum of the instrument.

$$\text{Tear Factor} = \frac{\text{Tearing Resistance in grams}}{\text{Basis weight in GSM}}$$

(GSM = Grams/sq meter)

9. Impact Strength Test for Plastics

These tests are designed to measure the ability of the films to withstand fracture by shock The test is a measure of toughness of the material. It is a combination of deformation and breaking properties.

10. The Abrasion Resistance

This test is designed to measure the ability to withstand surface wear and rubbing. It is a measure of some mechanical properties like hard resilience. The procedure consists of abrading the sample with a wheel of standard abradent for a definite number of revolutions and finding the weight loss of the sample.

11. Heat Seal Strength

The test is used for the heat sealable plastic packaging materials. The heat seal strength may be expressed as percentage of tensile strength of the base material (gm/cm width). The strength of heat seal depends upon temperature, dwell time, pressure and the type of heat sealing surfaces and each material has optimum values under these conditions.

12. Environment Stress Cracking

The purpose of the test is to study the influence of some reagents like soaps, wetting agents, oils or detergents on plastics, determined by exposing the specimens for a specific time to those environments and observing the cracks.

13. Sorption Behaviour

Role of Water in Foods

Water is the most abundant individual constituent by weight in the majority of foods, water exerts an important influence on many aspects of food quality. Foods are often divided into 3 major categories according to the proportion of water they contain.

Dried Foods

Solids–where moisture is less than 11 per cent

Powder–where moisture is less than 5 per cent

Intermediate moisture foods like butter, jam, pet foods etc

Wet food–fresh fruits and vegetables.

Each category has got critical moisture content with the upper and lower limits within which the product is satisfactory. Eg. In biscuits the critical moisture level is 2 per cent, means above 2 per cent moisture level the biscuit will not be crispy or breakable form therefore it has to be maintained at 2 per cent level in order to maintain its quality.

Changes brought about by moisture

Physical

Physical stature of any product is mostly depending on the moisture content of the product, properties like hardness, free flowness, caking, lumping, crispness, softness are governed by the level of moisture within the product.

Chemical

Moisture brings changes like rancidity in fatty foods, putrefaction in proteins like that in cashew kernel.

Microbiological

Microbiological growth in any product is based on its water activity (a^w) and water activity is defined as the ratio of the water vapour pressure of the water vapour present in the food to the vapour pressure of pure water under the same conditions of pressure and temperature.

Water content of a food at a particular humidity is dependent on its water soluble constituents and the presence of colloids. When moisture content of a food is plotted against equilibrium relative humidity (ERH) on water activity at a constant temperature, the curve so obtained is called *sorption isotherm*. Isotherm curves of different foods vary both in shape and water present at each RH. Water content of a food at a particular humidity is dependent on its water soluble constituents and the presence of colloids.

Moisture isotherm is the most useful representation of moisture relations and generally gives all the information necessary for determining the degree of protection required for the product and the extent of barrier property required for the packaging material.

The isotherm curve provided here, describes the hygroscopic nature of the product 'D' value (danger point) is 4.4 per cent whose corresponding (ERH) is 23 per cent RH, and critical point (c) is 1.5 per cent whose corresponding ERH is 25 per cent. It explains that when the product is exposed to different relative humidities, the product doesn't have any change up to an ERH of 23 per cent, but at RH 23 per cent and above the product tends to change its physical nature from free flowing to slightly caking, colour and flavour also tends to change and when the exposed RH is around 25 per cent and above it is the critical point for the product whose corresponding Equilibrium Moisture Content (EMC) is 51 per cent. Therefore it infers that the product as a spray dried free flowing powder, must be protected with a moisture proof packaging material.

How to Do the Study?

A set of desiccators having different RH per cent can be arranged by taking concentrated sulphuric acid of different normalities as shown below.

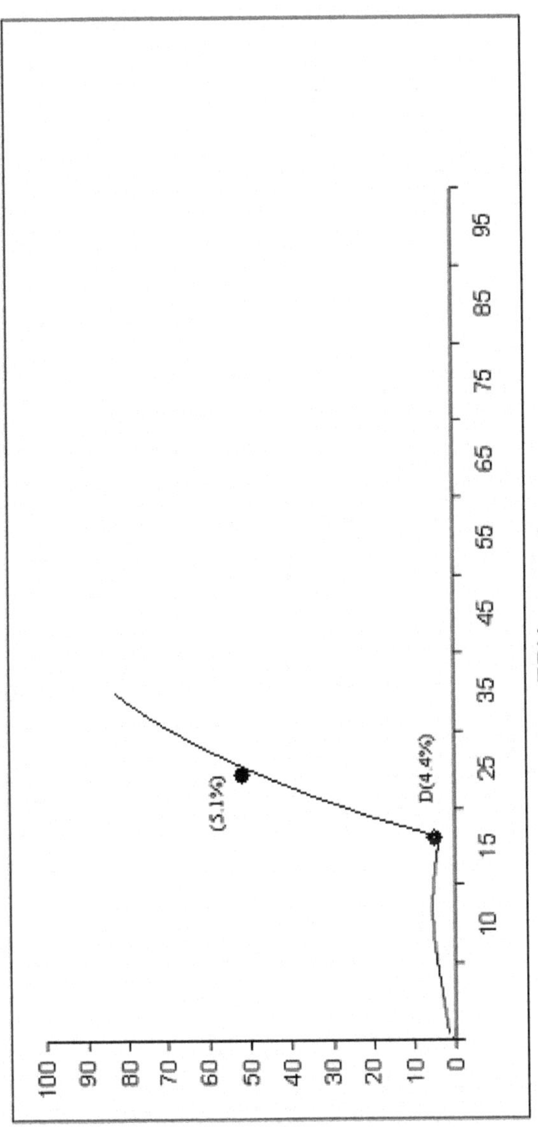

A Typical Isotherm Curve

D: Danger point (4.4 per cent); C: Critical point (5.1 per cent)

Normality H_2SO_4	Percentage RH
22	5
18	15
15.8	25
13.9	35
12.3	45
9.2	65
5.2	85
2.3	95

Into each desiccators weighed samples of the product in petridishes can be placed and allow it to remain for a period of 24 hrs at a constant temperature and after 24 hrs these samples can be weighed again and find the difference in weight with that of original weight in percentage (EMC). These values can be plotted against its corresponding ERH and the curve can be developed. Shape of curve indicates the nature of the product. A sharp increase in weight with respect to increase in RH shows that the product is hygroscopic and needs barrier proof material to protect the product from moisture absorption.

References

Abhijit Kar (2005) Quality testing of packages and packaging materials in the lecture schedule for diploma course in Food Processing and Engineering, IGNOU, New Delhi.

Baldev Raj(1995) Quality testing of packaging material and packages in the book Profile on Food Packaging Published from CFTRI Mysore.

Chapter 5
General Characteristics and Properties of Packaging Materials

I. Plastics

Plastics are organic polymers which are long chain molecules obtained by addition or condensation of one or more monomers. Polymerization of single repeating unit gives homo polymers and addition of more than one monomer results in co polymer. Most of the plastic used in food packaging are thermo plastic, means they can be softened by heating and hardened by cooling. Thermo plastic can be considered chemically as derivatives of ethylene ($CH_2 = CH_2$) in which one hydrogenation is replaced by CH_3, chlorine, fluorine or phenyl groups. These are often called vinyl plastics as they contain vinyl group

(CH_2 = CH) or poly olefins, since the monomer contain olefin linkage. Some of the polymers used in food packaging are:

General Advantages

- ✩ Wide range of barrier property to water vapour and gases
- ✩ Low density means light weight.
- ✩ Good strength
- ✩ Design flexibility
- ✩ Resistance to breakage
- ✩ Machinability–high speed filling using form fill and seal techniques or as pre made packs using variable sealing processes.
- ✩ Glossy and transparent
- ✩ Insitue colouring is possible
- ✩ High tensile strength
- ✩ High tear strength
- ✩ High clarity
- ✩ High printability
- ✩ High level lamination
- ✩ Cost performance ratio is very high
- ✩ Low storage and handling cost.

Disadvantage

The greatest disadvantage is the disposability means difficulty to get it disintegrated into soil.

Low Density Polyethylene (LDPE)

Most widely used material obtained by addition polymerization of ethylene gas under high pressure of 1000-

3000 atmospheres. LDPE is fairly soft, slightly translucent flexible material with waxy feel. It possesses excellent resistance to most chemicals. Good barrier to water vapour but less so to oxygen; has high permeability to volatile oils and swells in contact with fats and oil. Its low softening point prevents it being steam sterilized and in presence of some polar chemicals, it may subject to stress cracking. It gives very good heat seals and easily coated to other materials and serves as a laminated layer. It is used as bags, liners, bottles etc.

Properties

- ☆ Density ranges from 0.910–0.925 gms /cc
- ☆ Average molecular weight is 3×10^5
- ☆ Resistance to heat (82–100°C)
- ☆ Water absorption is 0.015 per cent
- ☆ Permeability to gases is $1.0cc/m^2/24$ hrs at 27^0C at one atmospheric pressure
- ☆ Good tensile strength properties and high percentage of elongation
- ☆ Resistance to weak acids and alkalis

Linear Low Density Polyethylene (LLDPE)

By low temperature and pressure of polymerization and use of Philips or co-ordination catalyst more linear polyethylene can be processed known as linear low density polyethylene (LLDPE)

Properties

- ☆ High tensile strength.
- ☆ High percentage of elongation
- ☆ High tear strength properties

☆ Better stress crack resistance and low temperature brittleness

☆ Improved stiffness properties

☆ Excellent puncture resistance

☆ Excellent heat seal properties

☆ Mostly used for the fabrication of bags, liners, and carry bags.

Medium Density Polyethylene (MDPE) Film

☆ Density varies from 0.926–0.940 gms/cc

☆ Average molecular weight is 2×10^5

☆ Resistance to heat (105–121°C)

☆ Translucent type of clarity

☆ Percentage of water absorption is 0.01 per cent

☆ Permeability to gases is $1.33cc/m^2/24$ hrs at 27°C at one atmospheric pressure

☆ Very resistant to weak acids and alkalis

☆ Effect of sunlight is yellowing.

High Density Polyethylene (HDPE) Film

HDPE is produced by low temperature 50–70°C and pressure (10 atmosphere) process using Ziegler catalysts and less waxy feel. It is stiffer, harder, less transparent and waxy feel. It is having better resistance to oil and greases, has higher softening point., but lower impact strength. Better water vapour and gas barrier property.

Properties

☆ Density varies from 0.941–0.965 gms/cc

☆ Average molecular weight is 1.25×10^5

- ☆ Resistance to heat (121°C)

- ☆ Opaque in nature

- ☆ High barrier to moisture vapour

- ☆ Permeability to gases is less compared to other polyethylene films

- ☆ Effect of sunlight is yellow.

- ☆ It is widely used for the fabrication of bottles bags, crates, tray etc.

High Molecular-High Density Polyethylene Film. (HM-HDPE)

By increasing the molecular weight of high density polyethylene, a paper like films with improved toughness can be obtained.

- ☆ High mechanical strength in both directions

- ☆ Pleasant white translucence in clarity

- ☆ High tear resistance property

- ☆ Dose not impart any taste or order

- ☆ Suitable for food contact application

- ☆ Less elongation as compared to other poly ethylene film

- ☆ Excellent moisture barrier property

Polypropylene (PP)

By using stereospecific catalysts and low pressures, isotactic form of polypropylene (PP) in which the CH_3 groups are largely ordered to the same side of the polymer chain is obtained. This isotactic polymer has higher crystallinity, is harder and has higher softening point than

HDPE. It is not subjected to environmental stress cracking, possesses good resistance to most chemicals and has permeability intermediate to LDPE and HDPE.

Properties

- ☆ High tensile strength
- ☆ High chemical resistance and high temperature performance than HDPE
- ☆ Very low permeability to moisture vapour and gas
- ☆ High transparency
- ☆ Chemical inertness
- ☆ High softening point.

Cast Polypropylene (CPP)

This film is widely used in food packaging, process of stretching the film in both directions under suitable temperature results in Biaxially oriented film (BOPP). It has improved glow, clarity, impact strength and better barrier properties to water vapour and oxygen.

It is mainly used in packaging as alternative to cellulose films for wrapping of snack foods and biscuits. It is extensively used as pouches, bags, liners, bottles, drums, snap on caps, woven sacks etc,.

Ethylene-vinyl Acetate Co-Polymer (EVA)

Contains up to 20 per cent of Vinyl Acetate and lesser properties similar to LDPE. It is more transparent and has high impact resistance, but higher permeability to water vapour and gasses. It is used as bags for stretch wrapping of frozen poultry and laminating due to good adhesion.

Ethylene Vinyl Alcohol (EVAL)

Ethylene vinyl alcohol is a material with extremely good gas barrier properties, unfortunately marred by a high sensitivity to water vapour, used as a middle layer in multilayer films for getting good gas barrier and better adhesion properties.

Poly Vinyl Chloride (PVC)

This polymer is obtained by polymerization of vinyl chloride monomer (VCM) and is used in two distinct forms un plasticized form as a rigid sheet suitable for thermoforming and plasticized form which is flexible.The presence of chlorine atoms in the polymer chain increases the polarity and formation of stronger inter chain forces. The plasticizers (usually aromatic ester) soften the resulting film and makes it more flexible but lowers the tensile strength. The film is widely used in the biaxially oriented form for shrink wrapping of meat and cheese. PVC film is suitable for thermo forming which provides, thermo formed blister packs, it is also used for injection moulding and blow moulding.

Properties

- ☆ It is hard, brittle and transparent material
- ☆ Low GTR
- ☆ Moderate WVTR and good resistance to fats and oils
- ☆ Glass like clarity
- ☆ Good mechanical strength
- ☆ Excellent printability

☆ Lower weight/volume ratio

☆ Resistant to chemicals

☆ Suitable for packaging carbonated drinks, mineral water and cooking oil. Highly suitable for cling or stretch wrap applications.

Polyvinylidene Chloride (PVDC)

Usually co-polymer of PVC and PVDC (saran or cryovac) is used as a self supporting film giving very good barrier and strength properties. Co-polymer is used for coating on paper, cellulose film, PP and also rigid plastics when very good barrier property is required. Shrinkable PVDC is used as counter wrap for poultry, meat and frozen foods.

Poly Tetra Fluro Ethylene (PTFE)

Here all the hydrogen atoms in the ethylene molecule is fully replaced by fluorine atoms. The resulting polymer is highly crystalline with very high molecular weight. Due to very high carbon-carbon bonds and carbon- fluorine bonds the polymer is stiff and strong. It is smooth waxy and usually grey in colour. Due to low coefficient of friction it gives "non-stick" surface capable of withstanding wide range of temperatures mainly used as coatings on frying pans and other cooking ware.

Polystyrene (PS)

Polystyrene consists of an ethylene molecule in which one of the hydrogen atom has been replaced by phenyl group. PS is a rigid, brittle, transparent plastic. It is fairly permeable to gases but has high WVTR. To reduce brittleness, it is often co-polymerized with acrylonitrile and butadiene to give high impact polystyrene (HIPS). PS is a versatile material for injection moulding and thermoforming as trays and tubs for dairy products and also used in packaging of fruits and vegetables and also as refrigeration lining. It is brittle unless biaxially oriented.

Condensation Polymer

Polyethyleneterephthalate (PET)

It is a condensation polymer and oriented heat set form. It is thermally stable and transparent. It has high tensile strength, bursting strength, dimensional stability, aging characteristics and gloss. It has low gas and volatile matter permeability but moderate WVTR. It resists oils, fats, dilute acids and also high temperature. Its availability and ability to be handled in thin gauges (12 micron) makes it suitable for lamination and vacuum metallization. Metallization of polyester film with aluminum reduces considerably WVTR and GTR.

Properties

- ☆ It has got excellent gloss
- ☆ Very low moisture and gas permeability
- ☆ High mechanical strength
- ☆ Resistant to tear, puncture, burst and flex crack
- ☆ Dimensionally stable over a wide range of temperature from 70°C to 130°C
- ☆ Excellent machineability
- ☆ Excellent printability
- ☆ Light in weight
- ☆ Free from all kinds of additives

☆ Good surface properties for metallization.

☆ Polyester bottles which are made by stretch blow moulding are increasingly being used to pack oils, liquors, sauces and carbonated beverages.

Polyamides (Nylon)

Polyamides are polymers obtained by the condensation of a diacid and diamine. Nylon-6,6 is made from adipic acid and hexa methylene diamine.Nylon 6, which is commonly used in food packaging is prepared by self condensation of one molecule ie ω-amino caproic acid. It is having high mechanical strength, high elongation capacity, excellent resistance to cutting, perforation, abrasion and bursting. High chemical resistance to oils and fats. Outstanding impermeability to gases and vapour. Easy to print, easiness in metallization and economical in usage. They could be biaxially oriented to make self standing pouches. Because of its high temperature withstanding property it is used in making boil – in- bag or roast-in-bag applications for convenient foods. Nylon is used as a core layer in co-extruded films used in packaging of oils and fats.

Modified Cellulose

These materials though derived from natural cellulose, are chemically modified into thermoplastics. The common ones are cellulose acetate, cellulose acetate –butyrate and cellulose acetate-propionate. These are used as thermoformed blisters, windows of modified atmosphere packages.

Regenerated Cellulose

These are not truly plastics but usually used in

combination with plastics. The cellulose from wood chips is taken into solution and reprecipitated as a continuous transparent film. The basic film is very good barrier to gases, but is sensitive and very permeable to water vapour. Coating with nitrocellulose or PVC protects the film as well as providing water vapour barrier and helps in heat sealing. Coated films like MST and MXXT are commonly used as they have less permeability, stiffness, ease of printing and provide good heat seal and fat, volatile and water vapour resistance.

Laminate

A laminate is defined as a combination of distinctly different flexible materials such as paper, cellophane plastic and foil where in each ply is generally thicker than 6 micron. Laminate films are bonded by process of dry bonding, wet bonding, and thermal extrusion and hot melt lamination. Laminate provide good barrier properties, heat sealability and functional properties. Examples of laminates are paper/foil/Poly, metallised polyester/Poly, cellophane/Poly, polyester/metallised BOPP, etc.

Co-extruded Film

Co-extruded films or multilayer films in which each distinct layer is formed by simultaneous extrusion process. Blow film extrusion is the most common technique with polyolefins. Combinations include, Polyolefins, Nylon, Ionomers, PVDC and Ethylene Vinyl Alcohol. Example-LD/LLDPE, LD/HD, LD/HD/LD, LD/BONDING/NYLON/BONDING/IONOMER, LD/BONDING/EVOH/BONDING/LD, EVA/BONDING/EVOH/BONDING/EVA

Three and five layers co extruded films have been extensively used in packaging of milk, fruit juices, oils and fats, cheese and meat products.

References

Anon (1985) PET its time has come. Food Engineering International.

Balasubramanyam,N.(1995) Plastics in food Packaging in the book Profile on Food Packaging, published from CFTRI, Mysore.

Barnetson, A.(1996) Plastic materials for packaging. Rapra Technologies Ltd.

Developments in Food Packaging.(1980) Edited by S.J. Palling Applied Science publishers Ltd., London.

Plastics in packaging(1984) published from Indian Institute of Packaging, Bombay.

Lamp,R.A.(1997) Flexible Packaging for thermo-processed foods. Advances in Food Research.

Mahadeviah, M.(1976) Plastic containers for processed food products. Indian Food Packer.

Plastics in Packaging(1984) published by Indian Institute of Packaging. Bombay

http://www.ecvm.org/code/page.cfmid-page:1.7

http://www.filmax.co.kr/eng/e-02.htm

http://enwikipedia.org/wiki/plastic

II. Metal

Metal containers are mostly used as rigid containers for the packaging of various preserved food products. In advanced countries about 50 per cent of the packaging materials are metal cans. Metal cans could be made either from aluminium, tin plate or tin free steel. But the tin plate containers and aluminium containers have got extensive application in packaging. The most popular form of metal container is tin plate container which has been used in food packaging for the last five decades. There are two types of containers namely.

☆ Open Top Sanitary Containers (OTS)

☆ General Line Container (GL)

These ranges of containers are standard variety containers from which a customer selects the size suitable for its requirement. These are also called as Open Top Sanitary cans (OTS) and are essentially round in shape. An open top sanitary can is available as its bottom often seamed to the body and the top ends are supplied as loose and sealed after filling in the processing factory.

Tin plate used for can making is a low carbon mild steel plate coated with a thin layer of tin. Steel base plate is manufactured either by hot rolling or by cold rolling process. Ninety percent of the steel used for tin plating is manufactured by cold rolling process. The coating is done either by hot dipping or by electroplating. Steel base plate consists of carbon, sulphur, phosphorous, copper, manganese, silicon etc of which phosphorus and silicon play a very important role in the corrosion of tin plate.

General Composition of Steel base plate

☆ Carbon 0.04 to 0.12 per cent

☆ Sulphur 0.015 to 0.05 per cent

Types of Steel Base Required for Different Products

Class of Foods	Characteristics	Typical Example	Steel Base Required
Most strongly corrosive	Highly or moderately acid product including dark coloured fruits and pickles	Apple juice, berries, Prunes, cherries, pickles	Type 'L' base plate
Moderately corrosive	Acidified vegetables	Sauerkraut	Type 'MS' base plate
Mildly corrosive	Mildly acid fruit product to low acid product	Apricot, figs, grape fruit, peaches	Type 'MR' or 'ML' base plate
Non-corrosive	Mostly dry and non processed products	Dehydrated soup, frozen foods and nuts	Type 'MR' or 'ML' base plate

☆ Phosphorus 0.015 to 0.06 per cent

☆ Copper 0.020 to 0.20 per cent

☆ Manganese 0.20 to.60 per cent

☆ Silicon trace to 0.80 per cent

The thickness of tin coating varies from 5.6g/sqm to 22.4g/sqm.Differential tinplate with different thickness of tin coating on either side are also available.

Advantages of Using Tin Containers

☆ High load bearing property

☆ Rigid

☆ Good tensile strength

☆ Perfect barrier if sealed properly

☆ Pilfer proofness

☆ Recylability

☆ Retortability under high temperature

☆ Amenable for high speed filling, sealing etc.

Disadvantages

☆ High tear weight

☆ Requires more number of joints and sealing.

☆ Amenable for corrosion

☆ Effectiveness of tin coating depends on its thickness and uniformity of coating

☆ Method of application of tin

☆ Composition of steel base plate

☆ Type of food and other functions

Various Types of Materials Used for Can Manufacture

Type of Steel Plate	Major Treatment	Type of Can
Mild steel plate	Principal material	Drums
Terrene plate	Mild steel coated with tin/lead	Drums
Galvanized steel	Steel coated with zinc	Drums and buckets
Aluminium	Pure aluminium or aluminium with 1 per cent Mn alloy	Foils and collapsible tubes
Stainless steel	Nickel-chromium coated steel	Drums, kegs or casks
Tin plate	Mild steel coated with tin	Cans and box
Tin free steel	With chromium and its oxide	Cans

Can Types Based on Closures

Type	Use	Particulars
Open Top Sanitary Can (OTS)	Fruits/vegetables/meat/ Fish products	Easy to fill and dispense solid products
Key opening cans with re closure facility	Coffee/cashew nuts etc	Subsequent re closure by a push in lid (frictional engagement)
Key opening can with no re closure facility	Canned products	Once opened no re closure facility
Slip lid cans	Dry powders	Simple re closure by pushing on the lid not suited for canning

Types of Can Closure Engagements with the Container

- ☆ Frictional engagement *e.g.* slip in/on lids
- ☆ Screw thread engagement *e.g.* pilfer proof caps
- ☆ Permanent mechanical interlocking *e.g.* can ends
- ☆ Vacuum closures due to internal vacuum.

Types of Cans

Three Piece Cans

Consists of a body and two ends, in the beginning food was filled into cans through a small hole in one of the ends and after processing, the hole was closed by a disc or cap by soldering. This was called hole and cap containers, but they had some disadvantages like difficulties in packing large pieces of food and difficulties in cleaning before filling. Later on these cans have been replaced by open top sanitary cans (OTS) which consists of cylindrical body with a soldered lock seam and un soldered double seam ends. Precision should be maintained in fabrication and seaming of OTS cans to achieve hermetic seal.

Flow Chart of Fabrication of Three Piece Cans

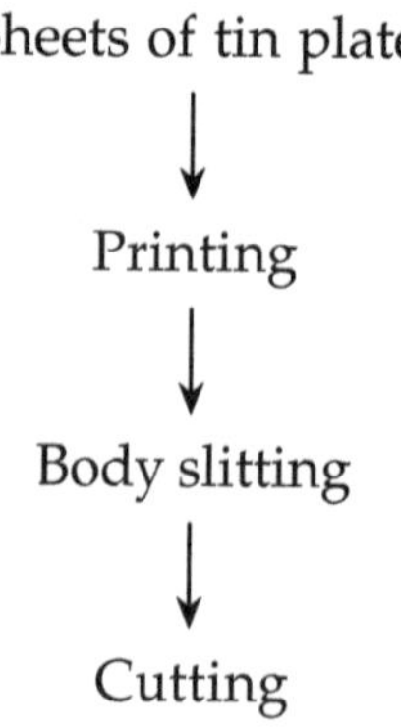

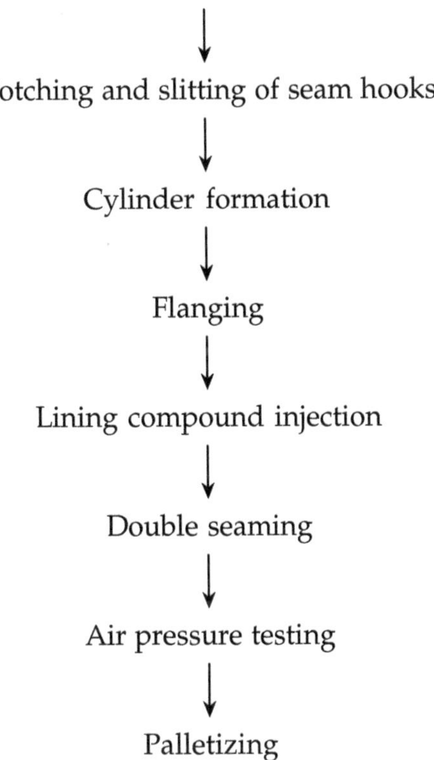

Two Piece Cans

Consists of only two pieces and mostly used for packaging beer and carbonated beverages where they need to have protection against the internal pressure due to carbonation. There are two types of two piece cans they are

Shallow Drawn Cans

These cans are extensively used for packaging solid processed products like fish, meat, fruits and vegetable products, so that dispensing the product will be easy and the containers will be with wide mouth and shallow depth.

Shallow drawn two piece can with easy open end (EOE) lid

**Shallow Drawn Aluminum Can with
Wide Mouth to Fill Solid Products**

Deep Drawn Cans

These cans are used for beverages and beer where the height of the can is more than the diameter and sleek in shape with easy end opening facility to hold the can easily and to draw the product. Cans with maximum height of

1.2 times the diameter with maximum diameter of 85mm is the most common one.

Impact Extruded Cans

The impact extrusion process is used at present to produce aluminum collapsible tube and aluminum

Deep drawn aluminum cans with liquid products

aerosol containers. Collapsible tubes are flexible containers for the storage and dispensing of product formulations which have pasty consistency. It consists of a cylindrical container with a shoulder, nozzle and closure at one end. The other end is open to allow the filling of the product before it is finally crimped to provide a completely sealed hygienic dispensing pack.

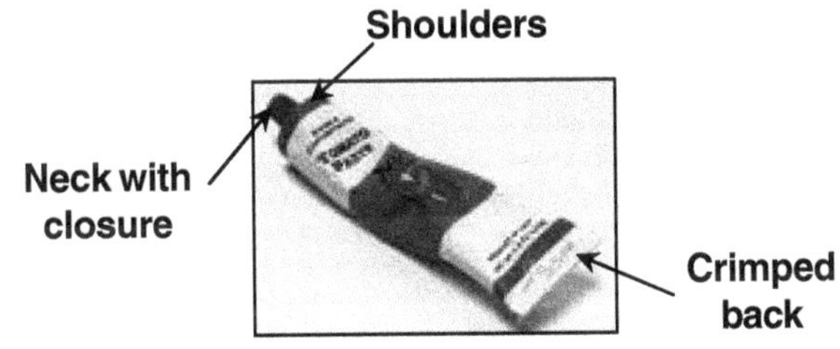

Flow Chart of Manufacturing of Two Piece Cans

Coils of metal sheet

↓

Cupping

↓

Ironing process

↓

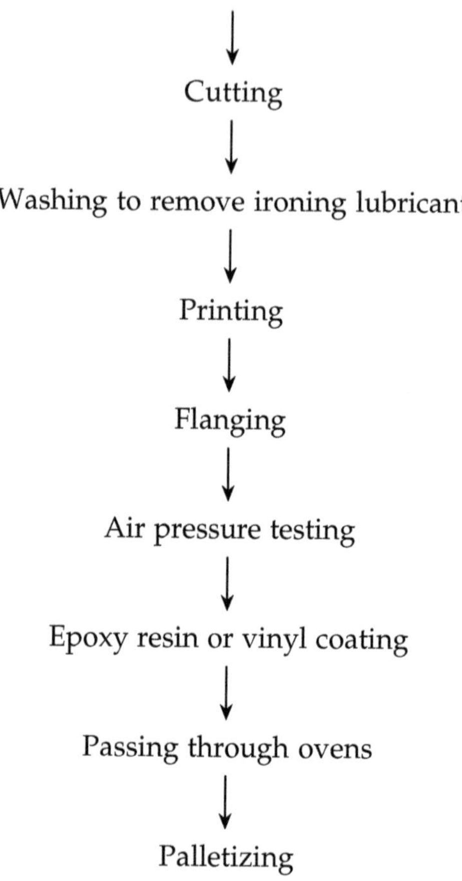

Easy Open Ends

Easy open ends have been introduced for beverages and fruit juice cans as a convenient device to open the cans. Initially a tear off tongue with a special incision and can end is removed as ribbon of metal. The latest type has a circular incision or a triangle incision which can be removed while pulling an attached ring. This is ideal for beverages.

Types of Cans Based on Method of Manufacture

- ☆ Seamless body type cans
- ☆ Usually pressed into the requisite shape and has no joints. It has no large surface area relative to depth. Lid will be mostly slip on type.
- ☆ Locked corner body
- ☆ Has an almost cubical shape
- ☆ Rolled body
- ☆ Has basically two components, a base and a side wall.

Corrosion of Tin Plate Containers

Corrosion of tin containers may be due to the influence of constituents of food product, processing variables and tin plate variables.

Influence of Constituents

The constituents of food products such as organic acids, anthocyanin pigments, polyphenols like rutin, gallic acid, catechin and pyrocatechol, degradation products such as hydroxymethyl furfurol, dehydro ascorbic acid, deketogluconic acid and demethylated pectin, polyphosphates and other complex agents like EDTA and sodium citrate are known to act as corrosion accelerators. Mango nectar prepared from unpeeled mangoes gave more corrosion which was attributed to the presence of gallic acid.Among the organic acids oxalic acid and oxalates are considered as detinners. The azo type of food dyes such as amaranth, tartrazine, sunset yellow, ponceau, etc used in low pH (2.5 to 4.0) carbonated beverage was found to accelerate corrosion. Nitrate present in some fruits and vegetables such as papaya, ivy gourd, tomato, carrot, green beans etc are considered to be the potential accelerators of corrosion. Such products may be packed in suitable lacquered cans.

Effect of Processing Variables

Processing variables which accelerates corrosion of tin plate are more head space, oxygen, low vacuum, in sufficient exhaust, low filling temperature,processing time and temperature, time lag between sealing and processing, high cooling water temperature and storage conditions.

Effect of Tin Plate Variables

Among the composition of base plate sulphur, copper and phosphorous affect the rate of corrosion. Hot dipped tinplates with phosphorus content of 0.02 and 0.025 per cent in base plate were found suitable for canning fruit and vegetable products, where as tinplate with phosphorus content of 0.03 per cent was suitable for canning mildly corrosive products like vegetables.

Effect of Corrosion

As a result of corrosion of tin plate, dissolution of metal like tin and iron occurs and results in the evolution of hydrogen which causes the ends of the cans to swell. Excess of metallic contamination imparts metallic taste to the product and in some cases affects the colour also.

Toxicity of Tin

Although tin is not considered as a poisonous metal, the presence of large dose produce serious digestive disturbance. Most of the tin present in canned food is insoluble in gastric and intestinal juices and is not absorbed during the process of digestion. In solid foods tin is mostly protein bound and in low concentration it has no significant toxic effect.

Inhibition of Corrosion

Corrosion of food containers can be inhibited by incorporating inhibitors to the product or by using suitable lacquered cans or by giving passivation treatment to the tin plate.

Inhibitors

The possible inhibitors tried to inhibit corrosion with different products are thickening agents like gelatin, agar-agar, pectin, carboxy methyl cellulose etc. Cysteine was found to reduce corrosion in soft drink cans and mango nectar.

Lacquers

Lacquered cans are used for packing few products to prevent corrosion and to protect the coloured fruit products containing soluble anthocyanin pigments.

Passivation Treatments

In recent years the importance of passivation treatment has been recognized in reducing corrosion of tin plate. Various combinations of sodium dichromate, sodium chromate, sodium phosphate, sodium hydroxide and sodium sulphate have been tried as immersion solutions for passivation treatment. Although electrolytic tin plate is already given light passivation treatment during manufacture to prevent rusting during storage of tin plate, passivated tin plate is not being exclusively used commercially for reducing electro chemical reaction of tin plate.

General Types of Can Coatings

Enamel	Typical Uses	Type of Lacquer
Fruit enamel	Dark coloured berries, cherries and other fruits requiring protection from metallic salts	Oleo resinous lacquer
Sulphur enamel	Corn, peas and other sulphur bearing products	Oleo resinous with suspended zinc oxide pigments
Citrus enamel	Citrus products and concentrates	Modified oleo resinous compound
Sea food enamel	Fish products and meat spreads	Phenolic lacquer
Meat enamel	Meat and various specialty products	Modified Epons with aluminum pigments
Milk enamel	Milk, eggs and other dairy products	Epons

Precautions to be taken while Canning to Prevent Corrosion

☆ Ensure proper head space and good vacuum.

☆ Prevent damage to lacquer coating while handling and seaming.

☆ After filling, cans are to be dried perfectly.

☆ Cans has to be stored under cool and dry places.

☆ Adequate cross ventilation must be provided in the store house.

☆ Leaking containers must be promptly removed.

☆ Non-corrosive adhesives must be used for sticking labels.

☆ Use pallets for stacking the cans.

Tin Free Steel (TFS)

Since the ore containing tin occurs only in few places and it is often difficult to secure, it has become necessary to find alternative to tin plate containers. Chromium coated steel plate, which is popularly known as Tin Free Steel (TFS) is one of the alternatives.

Commercial development of chrome plated and chromate treated steels for food cans began in Japan, Europe and Britain. Commercially available chromed steel in USA has a chromium coating of 4-9mg of metal/sq.feet. with 2-4 mg of chromium as oxide in the passivation film.

Chromium steels have proved in many instances to be more economical than tin plate. These materials have an attractive appearance, give excellent lacquer adhesion and can be drawn like other steel base material. They are resistant to under film corrosion and sulphur staining. They are also tolerant to high temperature.

None of the chromed steels can be soldered, but techniques have been developed in USA for forming side seams by high speed pressure or forge welding or by using thermo plastic cement. Chrome plated steels also find application for crown seals and screw caps for general line cans. TFS cans are ideal for packing different types of food products like low acid vegetables, sulphur containing vegetables, meat and fish products, milk powder, edible oil, ghee, vanaspathi etc

Aluminium Cans

Aluminium being one of the most versatile media, has over the years found extensive applications in the field of packaging. It is used in different forms like collapsible tubes, containers, foils, caps and closures. It has been well established that aluminium foils, in the form of laminates have revolutionized the food packaging industry due to their best flexibility and barrier properties.

India posses 80 per cent of worlds bauxite ore and has the potential to become the World's major aluminium producer; together with abundant advantages like availability and recyclability, its corrosion resistance also gives it a great preference as a packaging material.

It has sveral advantages like lighter in weight and compatibility with many foods. Introduction of aluminium easy open ends in beverage cans has boosted the market for two piece cans. Over 90 per cent of aluminum cans manufactured are in the beverage industry.

In the chemical and food industry aluminium of 99.5 to 99.7 per cent purity with the addition of one or more elements like magnesium, silicon, manganese, zinc, copper etc.to obtain the desired composition are used. These alloys have good corrosion resistance with most foods. Therefore the use of aluminium containers has increased many folds.

References

Mahadeviah, M.(1970) The alternative to sanitary tin can for packing food products. Journal of Food Science and Technology 7(3).

Mahadeviah, M. (1976) Internal corrosion of tin plate containers with food products. Indian Food Packer (2).

Mahadeviah, M. Gowramma, R.V. and Naresh,R. (1985) Protective lacquers for processed food cans. A review in Indian Food Packer.

Mahadeviah, M. and Gowramma, R.V.(1995) Packaging of processed fruits and vegetable products; meat and marine products in the book Profile on Food Packaging published from CFTRI. Mysore.

Mohan Kuruvilla (1997) Tin plate marketing in India. Journal of Packaging Technology published from Bombay.

III. Paper and Paper Related Materials

History of Paper

The term paper was derived from Greek word *"Papyrus"-an ancient* Egyptian writing material. Egyptians first used paper in 3500 B.C and in China by 2nd century from mulberry bark. The use of paper in Asia and Europe was started from 1310 and in America from 1690.

Advantages of Paper as a Packaging Material

Most widely used packaging material because of its stiffness and printability.

Properties of Paper as Packaging Material

- ☆ Good stiffness
- ☆ Good absorbent
- ☆ Low density.
- ☆ Not brittle
- ☆ Good creaseability
- ☆ Good printability
- ☆ Biodegradable
- ☆ Low cost

Disadvantages

- ☆ Poor tensile strength
- ☆ Poor wet strength
- ☆ Tear easily
- ☆ No barrier property without coating

Source and Manufacture of Paper

Paper is made from wood and wood consisting of

- ☆ Cellulose fibre 50 per cent
- ☆ Lignin 30 per cent
- ☆ Carbohydrate 16 per cent
- ☆ Proteins, resins and fats 4 per cent

Lignin holds the fibre together and by mechanical means the fibre bundles are broken and fibres are released.

Methods of Pulping

★ Mechanical pulping

★ Chemical pulping

Mechanical Pulping

Logs of wood chopped and ground with continuous supply of water, Soluble parts of wood will be lost through water washing, Lignin left after washing is not soluble in water but holds the cellulose fibres

Chemical Pulping

Removes everything except cellulose, obtain long cellulose fibres which has to be broken down into small fragments by continuous beating to give fibrils, these fibrils will be bleached continuously and such fine fibrils were bound together called fibrillation.

Chemical Processing

Kraft Process or Sulphate Process

Kraft means strength, therefore stronger paper is manufactured by Kraft process.

Wood chips + caustic soda+ sodium sulphate =cellulose fibres.

Sulphite Process

Less strong papers are manufactured by this process

Wood chips in aqueous solution of Sulphur dioxide + Calcium bisulphate and then heated to get cellulose fibres and are washed with water and further bleached with Calcium hypochlorite to get the sulphite paper.

Semi Chemical Process

Here wood chips are partly treated by chemical and partly mechanical means. Paper manufactured by semi chemical process is usually used for fluting medium for corrugated boards.

Properties of Paper

Physical Properties

Basis Weight

Fundamental properties of paper or paper board means weight of pulp used to make unit area of paper, expressed as gm/sq. m (GSM) or pounds/1000sq.ft.

Bulk

Volume or thickness of the material.

Decrease in bulk result in smooth, glossier, less opaque, darker paper but low in strength.

Friction

Resisting force between two papers when surfaces slide against each other measured as coefficient of friction. There are two types of friction static and kinetic. Static-force resists initial motion and the kinetic force resists motion of 2 surfaces sliding against each other. High coefficient of friction means more resistance for printing.

Moisture

Vary from 2 to 12 per cent depending upon relative humidity, type of pulp, degree of refining and chemicals used.

Smoothness

Property for printing

The term "Finish" describes smoothness of a paper, "glossy" means tendency of bags to slide when stacked'.

Temperature and humidity at the place of storage has to be controlled in order to maintain the above qualities.

Beating

Fibers separated through the chemical process are further subjected to beating to separate out fibrils. Various properties like tensile strength, burst strength and tear resistance of a paper usually depend upon the beating time.

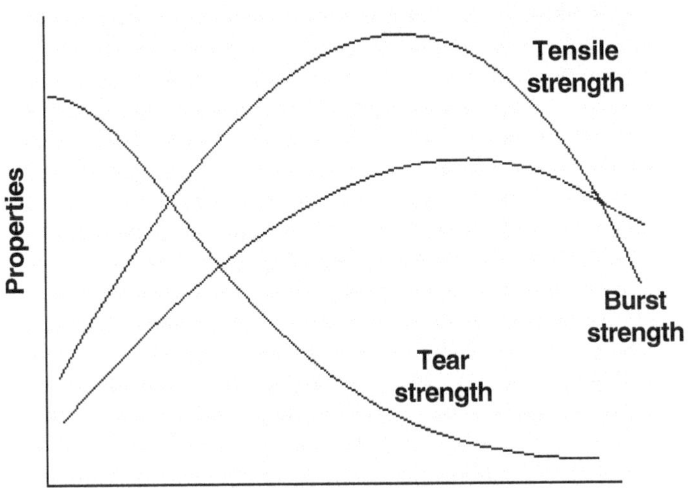

Tear strength decreases with increased beating time where as tensile strength increases with beating time, while burst strength increases with beating time only to a certain extent and then decreases with further beating. So depending upon the purposes of the paper, beating time can be varied to achieve the required paper quality.

Major Types of Paper Used in the Packaging Industry

Name	Sources	Uses
Kraft paper	Sulphate pulp from soft wood	Used as bleached, natural or coloured form, made wet strength or water repellent. Used for bags, multiwall sacs and liners for corrugated board. bleached form used for food packaging.
Sulphite paper	Made from a mixture of soft wood and hard wood pulp, usually bleached.	Clean bright paper of excellent printing nature used for smaller bags, pouches, envelops, waxed papers, labels and for foil laminates.
Grease proof paper	From heavily beaten pulp	Grease resistant for baked goods, industrial parts, protection for grease and fatty foods.
Glassine paper	Similar to grease proof but super calendared	Oil and grease resistant, odour barrier for lining bags, boxes etc for soaps, bandages and greasy foods.
Vegetable parchment paper (butter paper)	Treatment of unsized paper with conc. H_2SO_4	Non toxic, high wet strength, grease and oil resistant for wet greasy foods *e.g.* butter and fatty foods
Tissue paper	Light weight paper from moist pulps	Light weight soft wrapping for silver ware, jewelry, flower etc.

Beating will finally result in fibrillation; greater the degree of fibrillation, higher the strength of paper. Soft wood fibres fibrillate to a greater extent than hard wood fibers. Soft wood produces stronger paper.

Types of Carton Boards

☆ Lined chip board

☆ Unlined chip board

☆ Folding box board

☆ Fully bleached carton board

Corrugated Fibre Board Box (CFB)

The traditional packages are being slowly replaced by the use of corrugated fibre board boxes. These boxes are made from die cut corrugated fibre boards where the kraft papers are passed through corrugating machine to corrugated fibre board boxes. From another paper, prepare the fluting media or corrugation and finally stuck into a plain layer of Kraft paper by means of adhesives or gum to form 2 layer or 2 ply corrugation roll.

In the same manner, 3or 5 ply corrugated fibre board boxes can be made by pasting adequate number of Kraft liner or facing materials.

The fluting medium running in sinusoidal wave form between the two liners provides cushioning property. Depending upon the flute height, flute width and number of flutes per unit length will decide the extent of cushioning needed for the product to be protected. In the case of packaging of fruits and vegetables, ventilation holes for

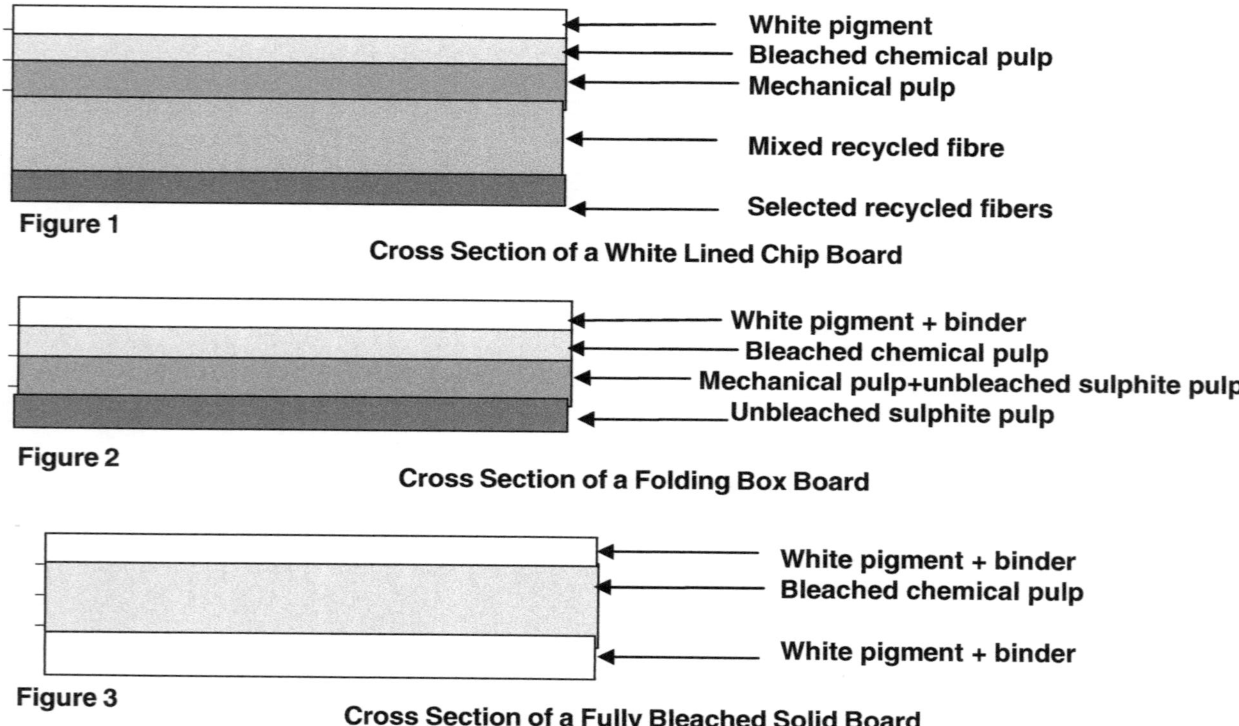

Figure 1

Cross Section of a White Lined Chip Board

Figure 2

Cross Section of a Folding Box Board

Figure 3

Cross Section of a Fully Bleached Solid Board

Board Type	Appearance	Uses
Unlined chip board	Grey fletched	Rigid and folding boxes for soaps, detergents, hard ware, electrical goods, boots and book binding
Brown lined chip board	Brown- grey body	Some rigid boxes in hard ware, electrical goods and shoe trades.
Kraft lined chip	Brown grey body	Mainly show card use
Cream lined chip	Cream/grey body	Used for folding boxes in food, pharmaceuticals and clothing industry
White line chip board	Off white liner with grey body	Used in cartons for break fast cereals, detergents, clothing, toilet tissues, hard wares, toys and games.

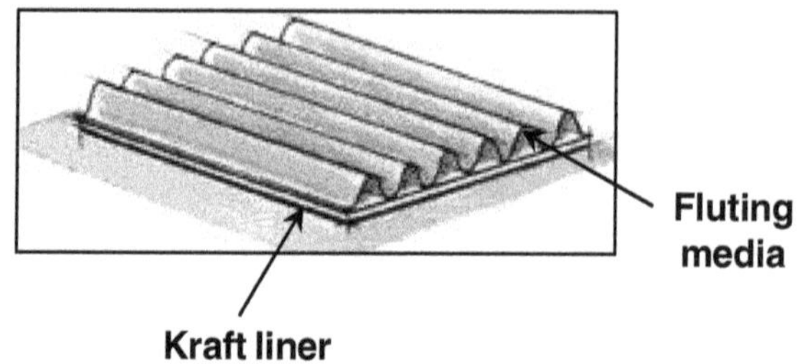

2 Ply Board (One liner + One fluting media)

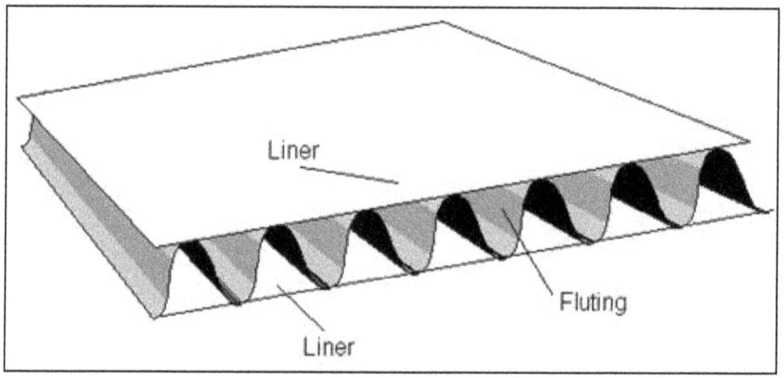

3 Ply Board (Two liners + One fluting)

dissipation of heat and exchange of gases are very much essential. Attention must be given to the number, size and shape and position of these holes without sacrificing the strength of the box.

Inserts and cushioning materials are generally used for packing glass bottles, pouches and fresh produce. Various types of inserts and cushioning materials that are being used are Cell pack–traditional partitioning method which contribute to the stacking strength of the box. Paper honey comb-can be adopted to different produces in different size and shape. Expanded poly styrene (thermo cole) can be produced in short runs at reasonable costs. Paper wool or wood wool- in combination with paper tissue gives good protection if tightly packed.

References

Development in food packaging (1980) Edited by S.J. Palling. Applied Science Published from London'.

Mahadeviah, M.(1985) Development in Food Packaging Material. Indian Food Industry 45.

www.paper board, boxes and cover-ebay.com.

www.sodaltech paper pulp moulding-paper plant.com

www.Food packaging and products.com

IV. Glass Containers

Glass is a senior member of the family of packaging materials and even today these containers are widely used in food packaging. It is defined as a neutral solution of silicates formed by heat and fusion, with subsequent cooling to prevent crystallization.

Advantages

- ☆ Provides perfect barrier
- ☆ Rigid
- ☆ Inert towards foods
- ☆ Transparent
- ☆ Good strength
- ☆ Can be moulded into any shape or size
- ☆ Can be recycled
- ☆ Colours can be added at the time of manufacture.

☆ Impermeable to gas or liquid

☆ Can be cleaned easily

☆ Glass containers are used for premium quality foods.

☆ Glass containers are distinctive, convenient and practical.

Disadvantages

☆ Brittle

☆ Subject to thermal shock

☆ High empty weight

☆ Needs separate closures

The main constituent of glass is sand, lime stone and soda ash. The sand used is known as silica or glass sand. Silica sand is melted at the first stage of making glass where soda ash helps to melt with greater ease. Glass made with the glass sand soda only will be partly soluble in water. When crushed lime stone is added, the resistance of glass increases greatly. There are few other chemicals that are added to these three main ingredients to make glass container stronger and cleaner or coloured as may be required.

Silicate may be replaced by borate or phosphate in certain cases for different purposes. For food containers silicates of Na and Ca with aluminum oxide, borate and barium silicate are mainly used. Mg oxide increase hardness of glass. Borosilicate have roughly $1/3^{rd}$ of heat expansion of ordinary glass, but has more resistance to shock and high electric conductance, excellent chemical stability and hence used for making kitchen utensils, laboratory glass wares, jars and bottles. Barium gives high refractive index, lead imparts weight and brilliance, zinc imparts toughness and heat resistance.

Colours are added to glass by incorporating suitable chemicals in small quantities as shown below:

- ☆ Ferrous iron: green colour
- ☆ Ferric iron: brown colour
- ☆ Chromium: green colour
- ☆ Manganese dioxide: pink colour
- ☆ Cobalt: blue colour
- ☆ Cadmium sulfide: yellow colour
- ☆ Uranium fluorescent: yellow colour

Common Ingredients for Glass Manufacture

1. Sand (SiO_2 silica): It exists as a polymer, $(SiO_2)n$.

2. Soda ash (sodium carbonate): Normally SiO_2 softens up to 2000°C, where it starts to degrade. adding soda will lower the melting point to 1000°C making it more manageable.

3. Limestone (calcium carbonate) also known as lime, calcium carbonate is found naturally as limestone, marble, or chalk.

The soda makes the glass water-soluble, soft and not very durable. Therefore lime is added increasing the hardness and chemical durability and providing insolubility of the materials.

Other materials and oxides can be added to increase properties (tinting, durability, etc.), produce different effects, colors, etc.

Main Properties of Glass

- ☆ Solid and hard material
- ☆ Fragile and easily breakable into sharp pieces
- ☆ Transparent to visible light
- ☆ Inert and biologically inactive material.
- ☆ 100 per cent recyclable

Manufacturing of Glass

Mixing-melting-forming and annealing are the four important operations involved in the manufacturing of glass.

Mixing

Depending upon the requirement of type and colour of glass containers to be manufactured, the raw materials are weighed and mixed thoroughly. For alkali, alkaline metals are used; for carbonate, soda ash and limestone is used; boron in the form of borax, for aluminium as aluminium hydrate, silica sand as silicon dioxide are the major ingredients.

Melting

The mixture is fed into melting furnace by maintaining a temperature of 2700°F (1560°C).

The furnace is lined with refractory bricks and free clay to withstand the melting temperature. The melted glass is then passed into refining chamber of the furnace. The impurities are retained in the melting chamber and the purified molten glasses are passed through a channel in the furnace called "Throat". Glass passing through the "throat" enters into the refining zone where some bubbles, normally escaped through the throat are allowed to be fined or eliminated and then enters to the gob feeder. At the precise position, just above the forming machine, the glass is allowed to leave the furnace through a cylindrical hole to the bottom of the feeder and shears to form a gob.

Forming

Molten glass as gob flows into a very large revolving bowl from which it is sucked up into the metal mold of rapidly revolving machine. Molds are usually made of cast iron to be shaped into a semi finished container. It is then transformed into a second mold, called finished mold, whose internal cavity is accurately machined to correspond to the desired external shape of the container. The glass container as it emerges from the finished mold has a temperature about 800°C, but it cools and hardens quickly.

Annealing

Glass containers are then transferred to a tunnel like annealing with planned zones to cool slowly. This gradual and regulated cooling of the container eliminates the internal stress and as it emerges from the other end of the Lahr. These containers are inspected for any faults or defects before packed into corrugated fibre board boxes.

History

General archaeological evidence suggests that the first true glass was made in coastal north Syria, Mesopotamia or old kingdom of Egypt. Modern glass was originated in Alexandria, artisans created a sort of glass called "mosaic glass" in which slices of colored glass were used to create decorative patterns

Types of Glass and Market Application

Soda-lime Glass

Common commercial glass and less expensive,60-75 per cent silica, 12-18 per cent soda, and 5-12 per cent lime; smooth and nonporous surface,inert, resistant to chemical reactions from aqueous solutions.

Disadvantage of Soda Lime Glass

Not resistant to high temperatures and sudden thermal changes.

Uses

For making bottles, jars, everyday drinking glasses, and window glass.

Lead Glass

54-65 per cent SiO_2, 18-38 per cent lead oxide (PbO), 13-15 per cent soda (Na_2O) or potash (K_2), and various other oxides. In moderate amounts, lead increases durability. In high amounts it lowers the melting point and decreases the hardness giving a soft surface. In addition it has a high refractive index giving high brilliant glass.

Borosilicate Glass

Silica (70-80 per cent), boric oxide B_2O_3 (7-13 per cent) and smaller amounts of the alkaly (sodium and potassium oxides) such as 4-8 per cent of Na_2O and K_2O, and 2-7 per cent aluminum oxide (Al_2O_3).

Greater resistance to thermal changes and chemical corrosion

Uses

Industrial chemical process plants, in laboratories, in the pharmaceutical industry, in bulbs making for high-powered lamps, used in the manufacture of cooking plates and other heat-resistant products.

Silica Coated Plastic Packaging

- ☆ Appealing benefits
- ☆ Barrier protection
- ☆ Recyclable
- ☆ Retortable and non retortable
- ☆ Provide 3 to 20 times longer shelf life protection

Glass Container Shape

Shape of the container is determined by the nature of the product. Each product group is having a characteristic shape.

Liquid Product

Small diameter finishes for easy dispensing.

Solid Product

Large diameter finishes for easy removal of contents using spoon or ladles.

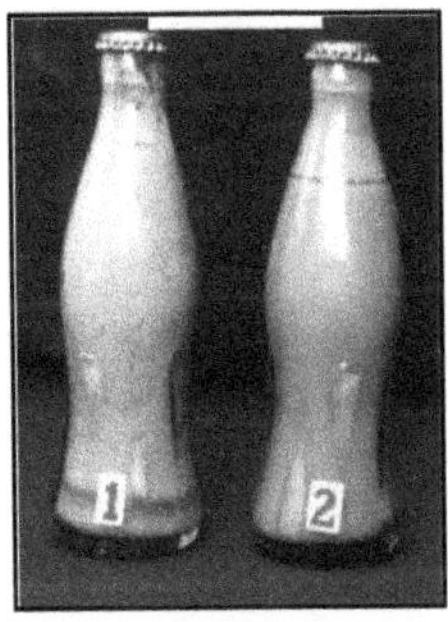

Narrow Necked Bottle with Metal Crown Cork for Liquid Products

Wide Mouthed Bottle with Metal Push on Lid

Closure (Caps) made Out from Metal or Plastics

Classified by,

Size – diameter

Sealing method – twist cap, cork

Special features – snap cap, pour out closure

Metal Closures

Sheets of tin plate and aluminium of 0.25mm thickness, coated with enamel to prevent the metal from reacting with the contents of the container is mostly used, the closure can be printed with graphics.

Plastic Closures

Generally either compression or injection moulded thermoplastic polymer like PS, LDPE, HDPE, PP and PVC are used for the manufacture of caps.

Requirement of Closures for the Glass Containers

1. It should prevent loss of the contents or any constituent of the content.

2. It should prevent penetration of any substance

3. The closure should not react with any contents.

4. It should be easy for the consumer to reach the content.

5. It may have to make a good reseal.

6. It must be pilfer proof

7. It should add sales appeal for the pack.

8. It should be rust resistant.

9. It should be economical.

Various Types of Caps

Crown Caps

It is applied by the pressure to the top of the bottle and seal hermetically into a glass ring built into the bottle neck. *e.g.* Beer, fruit drink and carbonated beverages.

Roll-on-Capping

Bottles to which this type of caps are applied have a simple continues thread built into the neck of the bottle at the time manufacture. *e.g.* jars for jams/pickles.

Push on Lids

This gives an air tight seal and is commercially used in capping bottles.

Twist On or Twist Off Caps

These are manufactured with a different thread design. Caps are brought preformed and generally contain a rubber or soft plastic seal inside. *e.g.* jam bottles/pickle, sauces, mayonnaise.

Corking

There are both natural and plastic corks available; natural corks are used to seal wine bottle. They can be used only once, while plastic corks can be used more than once.

Wine Bottle with Natural Cork

References

Mahadeviah (1985) Recent developments in food packaging materials. Indian Food Industry.

Moody, B.E. (1963) Packaging in glass. Hutchinson and Co. Publishers Ltd., London.

Osborn, D.G.(1980) Glass-Developments in Food Packaging Editor Palling, S.J. Applied Science Publishers, London.

http://www.enntech.com/glass.htm.

http:www.streetegyr.com/glass packaging industry market. Report.asp.

Chapter 6
Packaging of Different Products

I. Packaging Requirements of Spices and Spice Products

Spices are the aromatic plant materials in which aroma, flavour and colour are the main quality attributes. These three qualities attributes are highly sensitive to vagaries of climate and are prone to following deteriorations.

1. Loss of Aroma and Flavour

This loss in aroma and flavour is caused by the loss of volatile content due to seepage and oozing out through packaging material or due to oxidation of some aroma components.

2. Bleaching of Colour

This bleaching occurs in spices containing natural

colour pigments such as green pepper, green cardamom, turmeric, red chillies, capsicum, paprika and saffron and this loss is mainly due to the interference of light.

3.   Loss of Free Flowness

Spice powders become soggy and loose their free flow ness due to moisture ingress from the surroundings through the package. The extent of free flowness at different percentage moisture content, equilibrating to different relative humidities for different spices differ.

4.   Microbial Spoilage

When the moisture level increases in the product, the water activity will also increase. This will lead to the onset of microbial spoilage and depending upon the different levels of water activity, the type of micro organisms will also vary.

5.   Insect Infestation

This is caused by insects and rodents. These deteriorations are further favoured by humidity, oxygen, light and temperature.

Packaging Material Requirement

The packaging material intended for spices and their products should possess the following quality attributes.

1. Ability to protect the contents from spoilage and spillage.

   Packaging material must offer protection against physico-chemical and microbiological spoilage due to environmental conditions like humidity, temperature, light and oxygen. Therefore the packaging material must have a strong barrier proofness to water vapour and gases.

2. Material should have strong barrier proofness to aroma loss as well as to prevent the pick up of foreign odour.

3. Good oil and fat resistance characteristics.

4. Good machinability characteristics and should possess all required physical strength properties.

5. Should be good insect and mite resistant material.

6. Should be compatible with the product and confirm to the food laws of both exporting and importing countries.

7. Should have good appearance and printability to assist in selling by suitable attractive graphics.

8. Should be economical.

Packaging of Whole Spices

In whole spices, the flavour moiety is well protected by the cells from evaporation and oxidation; more over the surface area is also less when compared to ground spices. Therefore, the chances for the loss of flavour is less or negligible. Hence, the packaging material generally intended for bulk transportation to prevent, from moisture ingress in humid regions and from insect infestation in storage. The existing traditional package is the jute bag. A-twill, B-twill or DW gunny bags are used depending upon the value of spices. The weave clearance of these bags is 1-2 per cent, 3-5 per cent, and 4-6 per cent respectively which prevents spoilage and restricts movement of insects into the gunny bags. Sometimes double gunny bags are also used for better protection. Multiwall paper bags or cotton bags are also being used. However, in these jute bags, loss of volatile oils up to 23 per cent over a period of one year is observed. In order to avoid this loss polylined jute bags or HDPE woven sacks are being considered. Cardamom capsules which are green in colour may likely to get bleached are mostly packed in wooden or metal containers lined with a moisture barrier polyethylene material, *e.g.* 300 gauge black coloured polyethylene lined wooden chests. In case of dry whole chilly due to its low bulk density, volume poses a problem, hence compressed packs have been developed *e.g.* chillies with a moisture content of 10 per cent if compressed by applying a pressure of 2.5kg/sq.cm can reduce the volume by 78 per cent of its original volume with out causing any damage to the product. Dehydrated green and red pepper having moisture content of 3-4 per cent are very sensitive to oxidative changes like bleaching of colour especially in the presence of light. Hence need good protection against moisture, oxygen and light. Therefore plastic containers or poly bags of various sizes are being employed.

Packaging of Ground Spices and Spice Powder Mixes

Since the surface area of the particles get increased due to powdering, chances of moisture absorption and loss of flavour will be much more than that of whole spices, therefore deterioration is fast in such products. Moreover these products are intended for distant transportation and are sold in different markets as popular "masala mixes" both in internal and export market, needs both additional protection and sale appeal. A wide range of packaging materials are also available, depending upon the need of the manufacturer for quality and cost one can opt the material.

Poly olefins like LDPE, HDPE, PP, etc irrespective of their thickness, retards only the ingress of moisture and are inadequate in offering protection against loss of volatile oil. Further, printing on polyethylene pouches gradually get disfigured and smudged to great extent due to solubility and diffusion of organic volatile oils through these polymeric packaging materials. This is more predominant in the case of pepper and turmeric where volatile oil content is also higher.

Cellophanes and polyesters have less affinity for permeability of organic volatile oils and offer better protection against volatile oil loss. However, being hydrophilic in nature, have a high affinity to water vapour and are found inadequate for protection against moisture ingress under high humid conditions.Therefore a double pouch of cellophane inside and LDPE pouch out side will offer protection against volatile oil loss as well as moisture ingress.

Laminates of cellophane/poly, glassine/poly and 300 MXXT cellophane also retards volatile oil and prevents moisture ingress. How ever, these laminates are inadequate in providing protection against moisture ingress at very high humidity conditions.

Metallic polyester/poly and paper/Aluminium foil/ poly laminated pouches, which are very good barrier to moisture and volatile oils offer adequate protection for an year or over for different ground spices under Indian standard conditions. However, for pepper and turmeric powder which are rich in volatile oils, delamination of laminate was observed, to over come this problem a three ply laminate consisting of 50 gauge PET/paper/50 gauge metallised poly ester can be used. Presences of polyester or foil offers better resistance to insects than polyolefins and cellophanes and hence are better suited for long term storage. An outer duplex board carton offers further physical strength and good printing surface. These packages are well suited for export market and can offer one year shelf life.

Considering the economics for internal market either 200-300gauge HDPE or PP can be used for spice powders for short term storage. Laminated pouches of cellophane/poly, mettallised polyester/poly, paper/foil/poly are all suggested for long term storage of 200 days, half year and one year respectively under 65 per cent RH and 27°C.

Ground spices are bulk packed in multiwall paper sacks, textile and jute bags with moisture proof liners.

Packaging of Oleoresins and Volatile Oils

Pepper Oleoresin in Narrow Mouthed Glass Bottles

Oleoresins and volatile oils are obtained by extraction of ground dried spices and have great promise in export. This being highly volatile, have to be packed in tightly closed glass bottles, aluminium or suitably lined tin containers. These need to be protected from light and are to be stored in cool place. Aluminium containers with suitable inner coatings can be considered as consumer pack for oleoresins as it helps in easy dispensing.

One of the important problems in spice packaging relates to insect infestation during storage and distribution.

The extent of damage depends on factors like the initial contamination of the product, form and type of packaging material used and the type of insect itself.

All the flexible packaging materials have varying degree of insect resistance. It is observed that the polyester laminates offer maximum insect resistance among the common flexible packaging materials. Cellophane which allows some type of insects to penetrate within a week offers minimum insect resistance. Among polyolefins, PP is superior to polyethylene and HDPE offers better resistance than LDPE for insect penetration. Inclusions of duplex board carton to unit packs with suitable over wrap and proper sealing at the corners and proper fumigation will offer adequate protection against insect infestation.

Modern packaging materials such as coated BOPP (Biaxially Oriented Poly propylene) films, nylon and polyester based films etc. and special foil laminates in combination with cellophane, polyethylene, polyester, multilayer co extruded nylon based films etc. are used for packaging specialty spice products.

Multilayer films with different film structures have made significant in roads into food packaging and their use can be explored in view of their out standing barrier properties. Introduction of lined folding cartons and stand up packs based on polyester or nylon and polyethylene have significantly influenced the spice trade. They are light in weight, has desired barrier properties and are claimed to be cheaper than conventional metal and glass containers.

Vacuum and inert gas packaging of spice products in high barrier materials assures quality product even after one year of storage.

Retort Pouches for Wet Gravy Mixes

Retortable pouches are flexible packaging materials which has been developed as an alternative to tin plate cans in which low acid food products like meat,fish, vegetables and rice products which require high pressure processing at about 121°C is normally processed. Spice gravy mix which is a wet product unlike spice powder mix needs thermal processing to preserve without the involvement of chemical preservatives. Therefore the use of retortable pouches for serving instant spice gravy mix has become popular.

"Standi Pack" with Wet Gravy Mix Processed by Heat Sterilization

Retortable flexible pouches are laminated structures that are thermally processed like a metal can. They are self stable and have the convenience of Boil-in- Bag products, and have to provide superior barrier property, long shelf life, seal integrity, toughness, puncture resistance and must with stand the rigors of thermal processing. Most commonly used retort pouch consists of a aluminium foil and cast polypropylene. Polyester film is used for high temperature, toughness and printability; aluminium foil is barrier to light and gases. Cast poly propylene provides the critical heat seal integrity, flexibility, strength, taste and odour compatibility with a variety of food products and can withstand high processing temperature.

References

Anon (1981) "Sealed in flavours" Food flavourings, ingredients, processing and packaging 3(11) 23.

Balasubramanyam, N. Baldevraj, Indramma, A.R. and Andaswamy, B.(1980) Evaluation of polypropylene and other flexible materials for packaging of ground spice. Indian Spices 17(2).

Balasubramanyam,N. and Kumar, K.R.(1978) New concepts of making ground spices in flexible packages for internal and export market. Proceedings of Indian Convention of Food Science and Technologists.

Indramma, A.R.(1995) Packaging of spices and spice products. In the book profile on Food Packaging Published from CFTRI, Mysore.Indian Spices (1982) Vol. 13 (3 and 4).

Mishra. B.D.(1981) Storage studies on curry powder –effect of type of container, vacuumization, gas packaging and storage temperature on quality and shelf life of curry powders. Spice Bulletin 2 (6) 15.

II. Packaging of Cereal and Pulse Based Products

Cereal grains and pulse products constitute a major source of food and are chief raw material for developing many convenience foods. These products are highly sensitive to vagaries of climate and some factors which influence the quality are:

- ☆ Moisture
- ☆ Oxygen
- ☆ Temperature
- ☆ Humidity
- ☆ Respiration heat
- ☆ Biological factors like pest, rodents and microorganisms.

In order to protect the product from these factors a packaging material should have all the functional properties like

> (1) Ability to protect the content from spoilage and spillage
>
> (2) Should offer protection against environmental conditions and microorganisms.
>
> (3) Should posse's necessary strength properties to withstand mechanical hazards during transportation and storage.
>
> (4) Should prevent insect infestation and insect damage.
>
> (5) Should assist in selling by attractive graphics.
>
> (6) Must be economical and easily available.
>
> (7) Should confirm to the food laws of the importing and exporting countries.

The cereal and pulse based products can be classified as:

> (1) Whole grain product like rice, wheat, oats, millets etc.
>
> (2) Milled grain products like wheat flour, *atta, maida, rawa, soji, besan* etc.
>
> (3) Processed products like papads, vermicelli, break fast cereals, weaning foods etc.

Packaging Needs of Whole Grains

They are often packed as bulk package in sacks, multiwall paper sacks for commercial use. Most simple and common is the unlined jute bag of 100 kg capacity. They are having proper grip while stacking one over the other, but lack protection against insects and moisture. In certain cases in order to over come insect attack they have been treated with insecticides but the risk of grain contamination has retarded this development, where as in large go downs or ware houses, these bags are fumigated against insects. In order to protect against moisture absorption jute bags are lined with polyethylene bags. Recent years consumer units in super markets are ranging from 1kg to 25 kg for which multiwall paper sacks and plastic film bags are common. For unit packs of less than 2kg, plastic film bags of low and high density polyethylene and polypropylene of 50 to 75 micron thickness are useful as these offer desired protection against moisture; where as for export laminates of 12 micron over 50 micron polyethylene is a choice material.

Packaging Requirements of Flours

Most critical factor that deteriorates flours is the ingress of moisture leading to lumping, caking, rancidity, microbial spoilage etc. The deterioration is faster when moisture content exceeds 13 per cent. Proper packaging will help in minimizing the quality loss during storage and will also assure quality product under hygienic conditions. Early flours were dispensed in kraft or news paper bags, later it was switched over to cloth bags or textile sacks when quantified above 5kg packs. Jute sacks were also used for the bulk transportation of wheat flour, but sacks do not provide protection against the moisture ingress. Recently multiwall paper sacks and laminated/lined HDPE or PP woven sacks are being used to provide protection against sifting and moisture ingress.

In general for good storage stability 200-300 gauge (50-75 micron) low density polyethylene, 200 gauge (50micron) HDPE and 200 gauge (50micron) pp are quite suitable in offering desired protection against ingress of moisture under different climatic conditions.

Recommended Packaging Materials

1 kg Unit Packs of Atta, Sooji and Maida

(*a*) 200 gauge transparent polyethylene packs are the simplest and most economical one.

(*b*) Yellow coloured bright surfaced 350 gauge polyester/polyethylene laminated pack.

(*c*) Double pack of 150 gauge high molecular high density (HM-HDPE) film inner pack and outer white pigmented 230 gauge polyethylene.

5 kg Packs

(*a*) Double packs of inner 220 gauge HM-HDPE and outer yellow pigmented 320 gauge HDPE pack.

(*b*) White pigmented 450 gauge polyethylene bag with top handle for easy carrying.

10 kg Packs

(*a*) HDPE woven sacks stitched on two sides. This is an economical pack with protection against sifting and moisture ingress.

Basin

(*a*) Slight yellowish tinged opaque 280 gauge polyethylene of 500g pack.

(*b*) 200 to 300 gauge poly olefins such as low density polyethylene (LDPE) linear low density polyethylene (LLDPE), High density polyethylene (HDPE) high Molecular High Density polyethylene (HM-HDPE) and polypropylene(PP) offer good moisture protection at economical cost. They all are some what fairly resistant to insects and fumigants and can permeate through these films thus helping in insitue fumigation of packaged products. Impact resistance wise LDPE and LLDPE are very good compared to HDPE, HM-HDPE and PP.

Polyester/polyethylene laminate can offer protection even longer periods of storage due to good strength and barrier properties. Even though they are quite attractive, insitue fumigation is not possible, therefore products have to be fumigated prior to packaging.

For 2kg and above, use LDPE/LLDPE as it is better suited in view of its good impact strength and moisture barrier properties. For larger period of storage double packs consisting of HM-HDPE and LDPE are better suited.

Packaging of Processed Cereal and Pulse Products

Cereal flakes, like corn, maize, rice and wheat, vermicelli etc which are processed products are basically sensitive to moisture. Moisture ingress results in softening and microbial spoilage. Below certain lower levels of moisture they are fragile and brittle. Basically these products require moisture protection. Corn flakes are packed in duplex board carton over wrapped with cellophane. Products like puffed rice, rice flakes and maize flakes which are mostly packed by small cottage industries are often dispensed directly in news paper bags. Polyolefins in different thickness offer adequate protection against moisture. With the growing urban demand for pop corn, spice and sugar coated cereal puffs, it is better to be packaged with polyethylene, polypropylene, cellophane and for better shelf life and quality; laminated pouches of polyester/foil have better prospects.

Vertically Filled Packages in Lamininated Polymeric Package

Vermicelli

Traditionally made from wheat or rice is a popular item and is available in different types of attractive packs. Commonly used are 200 or 300gauge LDPE or HDPE. A modified form is 200gauge polypropylene.

Weaning Food

A product prepared from cereals and pulses are highly sensitive to moisture changes and become lumpy or caky above 60 per cent RH. Higher moisture levels of above 10 per cent in the product will lead to hydrolytic rancidity leading to bitterness.

Commercially weaning foods are generally packed in tin containers. Flexible packaging materials like polyethylene with high or low density and laminates based on paper/ foil/ poly or metallised polyester/poly can be considered as cost effective alternatives to tin containers.

Packaging Requirements of Typical Ready Mixes

Ready mixes like idli, dosa and cake mixes have very little fat and require moisture barrier rather than oxygen barrier packaging materials.At higher humidities they pick up moisture rapidly and a moisture content of 11-13 per cent are critical with respect to storage stability. Moisture barrier films based on polyolefins comprising of low and high density polyethylene, polypropylene (cast and BOPP) films of 50-75 microns thick are adequate in providing 90-120 days storage life.

For rich mixes like jamoon, cake, doughnut etc. which have high fat content and milk solids are highly prone to rancidity due to interaction of oxygen and moisture. For short period of storage life 50 micron cast polypropylene (CPP) is enough which can give a life of about 90 days. For longer shelf and export purpose plain printed polyester/poly laminate or foil laminates or co-extruded films based on nylon are better suited from the point of maximum protection and appearance.

Polypropylene packs are preferred to polyethylene even for short term storage as it provides better fat resistance, moisture and oxygen barrier properties. Two types of unit packs are common for all type of mixes, one commonly used being 50 micron polypropylene enclosed in attractive duplex board cartons. The other one is plain polyester/polyethylene laminated stand pack with attractive printing are the recent and convenient one most suited with an idea of export market.

References

Balasbramanyam, N.(1995) Packaging of cereals and pulse based products. In the book Profile on Food Packaging, CFTRI, Mysore.

III. Packaging of Dairy Products

The dairy industry in India is one of the largest among the food processing industries. Milk is one of the most important dairy product and other products are butter, cheese, ice cream, khova, curd, ghee, shrikhand etc. Packaging plays a vital role in preventing quality deterioration of milk and milk products. A functional package for dairy products in general has to protect the content against developments of rancidity, ingress of moisture, loss of odour,tainting and microbiological spoilage.

Packaging Materials for Milk and Other Products of Milk

Packaging materials selected should have the following requirements

(*a*) Prevent tainting problems caused by deterioration of fats present which will cause off odour.

(*b*) Prevent oxidation of milk fat which produces tallow flavour. This is accelerated by heat, acid and metal ions like copper and light.

(*c*) Needs protection from contamination.

Bulk movement of milk is done in tinned cans or aluminium cans of different capacities from 10 to 50 litres. Specification for insulated aluminium milk cans of 20, 30 and 35 litres for collection and distribution of fluid milk is given by IS 1825 in 1971. Chilled milk is transported from rural areas to large dairies in insulated tankers with capacities ranging from 1000 to 10000 litres.

In rural areas milk is sold as raw conditions by vendors who store and transport milk in aluminium cans. Since this

Aluminium Milk Cans of Different Capacities from 10 to 50 Litres for Short Distant Transportation of Fluid Milk and the Cans are of ISI Standards (IS1373) and (IS 4937)

cans are without sealing chances for mixing of water and contamination is a common draw back. Therefore, milk processed in large dairies for distribution is in packed and sealed consumer packs.

Earlier periods 500 ml glass bottles were very common in large dairy plants which were used as a multi trip package and on an average each bottle performs as many as 90 trips before it goes out of circulation. These bottles were amenable for machine filling and sealing at a speed of 90 bottles/minute and aluminium foil of 0.05mm thickness was used as the seal. These caps were not protected against puncture or penetration and since bottles were heavy, transportation cost were often high; besides it involves initial high capital investments for procuring bottles, needs large operational area and large store for empty bottles. However, these bottles were convenient, returnable and good insulating property and non-toxic.

Later bottles were replaced by flexible plastic pouches. These pouches are single trip and very light in weight and hence distribution becomes less costly when compared to bottles. Plastic pouches were formed, filled and sealed in a single machine called Form Fill and Seal (FFS) machine. Earlier single layer films of polyethylene were used and later this has been replaced by co-extruded multilayer films made of low density and linear low density polyethylene. These pouches are used in 500ml and 1000ml capacity with thickness of 65 and 75 microns respectively.

Pouches offer economy, compactness and ease of disposal. Disadvantages include the need for support and un conventional appearance. Since clear plastic does not offer adequate shelf-life an opaque co-extruded film such as white pigmented or LD/LLD pouches with inner layer as black and outer white in colour for flavoured milk is used. This prevents exposure of milk to light.

Tetra Pak

Aseptic packaging of milk in cartons was originally introduced in Sweden as early as 1960. "Tetra Pak" system consists of single trip containers made by laminates of high quality paper with foil and low density polyethylene. It is defined as a packaging of pre sterilized milk into a pre-sterilized package in an aseptic environment. "Tetra Pak" is sterilized using hydrogen peroxide. No refrigeration is needed and shelf life is between 60 to 90 days. Under typical Indian condition the shelf life is about 45days. Development of brick shaped "Tetra Brick Pak" has made a highly perishable product like milk into virtually a grocery product like selling and moving milk under ambient conditions.

Use of all plastic milk bottles which are tough and light in weight can be considered and the basic material used in many countries are polystyrene and poly carbonates; however these pose problems in capping and filling operations. Bulk packaging of 4 liters of milk in plastic containers have been introduced. Poly carbonate returnable bottles have received noticeable interest in developed countries. "Bag- in box" a new concept in recent years have become popular in many countries, it is a flexible bag made of mostly poly ethylene will be placed in a carton box as a secondary pack. 19 litres of bulk milk pack is available in such containers known as "liquid box' consists of 50 micron inner poly ethylene with a dispensing spout attached with the bag and this is having a valve to dispense the milk and the bag will collapse depending upon the use as the level of the milk falls; this whole inner pack will be placed into a carton box as a secondary pack to protect the inner primary pack, allowing the spout to be opened from outside the box.

Butter

Butter, another important dairy product is also sold widely; it consists of about 80 per cent milk fat and 15 per cent moisture and up to 3 per cent salt. Because of the higher percentage of moisture butter is always susceptible to mould growth. Presence of fat will lead to

Secondary Packaged Butter with Inner Parchment Paper and Outer Paper Chip Board Box

deterioration through rancidity or absorption of other materials. Mould growth can be inhibited by exclusion of air or by tight intimate wrapping by suitable packaging material; in order to prevent rancidity butter must be protected from light and oxygen. Therefore, packaging material intended for butter must posses resistance to fat, oxygen, moisture, foreign odour, light and must with stand refrigeration.

Most commonly used butter wrap is vegetable parchment paper; it is grease resistant but not sufficient enough to prevent oxygen ingress and its transparency allows some light. Hence for superior product protection and extended storage life laminates of aluminium foil are quite useful. Best is foil laminated with paper, foil offers opacity and barrier properties and paper gives strength, machineability and printing. Generally foil laminated with parchment paper or grease proof paper or tissue paper forms an ideal package for butter.

Butter can be sold as bulk in stainless steel or tin plate containers. For unit packs, tin containers are expensive therefore flexible packaging materials are common. Packs of 100 to 250gms are packed in parchment paper and placed in waxed cartons for refrigeration.

Cheese

One of the most sold dairy product with a thick competition existing in the world, therefore very many fancy packs of different shapes and sizes are available. The packaging material used for cheese must afford general protection, prevent moisture loss, prevent oxygen transmission, protect against micro organisms and must be fat and grease resistant. Oxygen can be eliminated by packing cheese in hermetically sealed containers in vacuum or in inert gas atmosphere. Processed cheese is usually packed in aluminium foil in cube forms with different sizes. Tin plate cans are used for 200g and above. PVDC coated plastic films are suitable as they provide good oxygen and moisture protection.

Cottage cheeses are often packed in rigid paper board cartons, but they have only short shelf life, therefore vacuum formed polystyrene tubs are used almost exclusively for packaging cheese. Plastic lined metal cans are used for special cheeses. Recently co extruded films containing nylon as core layer ie LD/nylon/LD has been employed to vacuum pack cheese. The use of nylon for lamination or for co-extrusion will offer excellent flex and abrasion; barrier and heat sealing property, an excellent combination of this kind will be a combination of PVDC and Nylon 6, which is having superior machineability and abrasion resistance.

Dried Milk Powders

Since whole milk powder and skim milk powder are rich in fat, packaging material should provide protection against oxygen and moisture. Generally in the beginning tin containers were used with the head space filled with nitrogen replacing oxygen, where as for skim milk powder, it contains only less fat, just moisture barrier material is adequate for short term storage. Skim milk powder can be sold as 1kg pack in poly ethylene bags or in HDPE plastic bottles for 200g and 400g powder.

For whole milk powder aluminium foil laminates like paper/polyethylene/foil/poly are excellent for retail packs. Vacuum packaging in oxygen impermeable flexible film laminates is better than tin containers with inert gas for whole milk powder in terms of shelf life and quality. For example 12 micron metallised polyester when laminated with 300 gauge polyethylene has proved to be better over tin containers with nitrogen flushing. Whole milk powder and infant foods require vacuum or gas packing to prevent oxidation of fat.

Ice-cream

As ice-cream is always stored and handled under low temperature the packaging material must have the low temperature stability. The package should need protection against contamination, attractiveness, ease of opening and reclosure

Ice-cream Cups made of High Impact Poly Styrene (HIPS)

and ease of disposal. Most important is the protection against moisture loss and temperature fluctuations.

Bulk ice-cream is packed in a linerless sulphite board carton, coated with wax or polyethylene wax blends. Generally paper board made of virgin chemical wood pulp coated with wax is used. Ice-cream bars are generally wrapped in vegetable parchment paper. High Impact Polystyrene (HIPS) cups are more popular now. Mostly for cups lid material is either wax coated paper board or plastic like LDPE or again with polystyrene. Plastic cups are moisture proof, leak proof and convenient to use. HIPS cups ranging from 5g to 35g with a thickness ranging from 0.3mm to 0.65 mm for capacities ranging from 100ml to 1000ml are available. Sweets like shrikhand, rasgullas, lassi etc are also packed in thermoformed containers made from HIPS. Lid materials of these containers may be either polycoated aluminium foil or polystyrene film. For short term storage lassi is packed in plastic pouches packed on FFS machines.

Ghee

Ghee is a product of 100 per cent fat, for long term storage stainless steel or tin plate containers are desirable. Multi layer co extruded films such as LDPE/HDPE are being used for short term storage. For long term packaging, material should have good fat resistance and barrier properties against oxygen and moisture. Metallised polyester/ poly pouches have the excellent gas and moisture barrier proofness similar to one packed in glass bottles, since it is flexible it can be preferred in terms of economy and convenience in packaging ghee.

Ghee needs protection from chemical spoilage to prevent rancidity activated by light, humidity and temperature. Besides ghee is likely to get off odour, picked up from the surrounding products. To prevent hydrolytic rancidity ghee needs fairly good moisture proof material. Therefore metallised polyester/polyethylene "stand packs" and lined folding cartons with paper/foil/poly laminate liner with good barrier properties have found to offer long shelf life. Ghee is also available in handy economical flexible sachets in 100g to 500g packs. Nylon based co extruded films of white pigmented polyethylene/nylon/Ethylene acrylic acid (EAA) having a thickness varying from 90 to 110 microns will offer a shelf life of 120 days in the normal conditions of storage. These flexible package of 200g sachet packed on FFS machine has really revolutionized the ghee packaging.

References

Abrahamsen, R.K. and Holmen, T.B. (1981). Packaging of dairy products. Jr. of Dairy Research.48, 457.

Adams,D.M. and Brawley, T.G. (1981) Packaging dairy products. Jr. of Dairy Science 58, 828.

Balasubramanyam,N.(1995) Packaging dairy products in the book Profile on Food Packaging published from CFTRI, Mysore.

Briston J.H. (1980) Developments in Food Packaging Edited by Palling S.J. Applied Science Publishers Ltd., London U.K.

Packaging of Food products (1986) Published by Indian Institute of Packaging, Andheri, Bombay.

Shashi Prabha, Vikas Gulia and Renu Dadarwal (2003). An innovative approach of food preservation. Indian Food Industry vol. 22.

IV. Packaging of Edible Oils and Fats

Oils and fats are chief components in human diet, due to their intrinsic favourable qualities of providing calories and physiological needs and extrinsic factors such as flavour, palatability, appearance, consistency and texture.

Oils are bulk packed in tin containers or galvanized iron drums and sold in retail. In order to avoid adulteration, easiness in handling and distribution and for the assurance of quality and quantity it has to be packed in unit containers of 1kg to 15kg in lined folding cartons, HDPE, PVC and PET bottles.

Nature and Deteriorative Characteristics of Fats and Oils

Both fats and oils have similar chemical composition in that they are mostly triglycerides of different fatty acids. The products which are solid or semisolid at ordinary temperature are called fats and those which are liquid are called oils. They are sensitive to climatic conditions and the deteriorations are mostly as follows:

Oxidative Rancidity

This is mostly caused by the oxidation of fats and oils and is promoted by temperature, humidity, UV light, trace metals and the availability of oxygen for reaction. Each oil has difference in the rate of oxidative deterioration due to their oxygen susceptibility. Oxygen susceptibility would be highest in case of oils containing higher degree of unsaturated fatty acids like oleic and linoleic acids and lowest in case of oils containing saturated fatty acids like palmitic, lauric and stearic acids under similar conditions. For hydrogenated fat like vanaspathi oxygen susceptibility is very low. Unrefined oils due to the presence of natural anti oxidants are less prone to oxidation than refined oils.

Hydrolytic Rancidity

This deterioration is caused by the hydrolysis of fat accelerated by the presence of moisture which is catalyzed by enzyme, this is common in fats and oils containing higher amount of saturated fatty acids. The rate of reaction is high when relative humidity around the product is more than 65 per cent corresponding to a percent moisture content of 0.10 per cent for refined oil. Therefore at higher RH oil require protection against moisture gain with perfect barrier proof material.

Tainting by Foreign Odour

Even though hydrogenated fat have greater stability than oils, it is equally susceptible to odour pick up. So the packaging requirements for the both needs excellent gas barrier material.

Microbiological Spoilage

When the moisture content rises it can lead to microbial spoilage, this is more accelerated when RH is above 65 per cent.

Colour Degradation

Colour degradation is further accelerated in the presence of light. Therefore packaging material needs light protection or opaque materials are to be used.

Packaging Material Requirements of Oils and Fats

To prevent or to slow down rancidity of oils and fats, for the easiness in handling and to offer protection the packaging material should posses the ability to maintain chemical quality, colour, flavour and other quality attributes.

- ☆ Packaging material should not impart any foreign flavour.
- ☆ Must be barrier proof to moisture, odour and oxygen.
- ☆ Must be resistant to grease.
- ☆ Must be opaque or pigmented to minimize U.V. light.
- ☆ Should not be toxic but must be compatible with product

☆ Should have good impact resistance to prevent product loss or contamination due to breakage or leakage of the package.

☆ Should posses stiffness, tensile strength and resistance to tearing as they are usually packed in automatic FFS machine for flexible films.

☆ Should be tamper proof and have air tight sealing.

☆ Should be economical, easily printable and easily disposable.

Packaging Materials

Flexible Pouches

Flexible packaging materials used to pack oils are generally co- extruded films of 90 to 100 micron thick pigmented white, yellow or green or laminates. The polyester and metallised polyester web used in conjunction with others will be generally 12 micron which provides the required barrier and strength properties. Examples are LDPE/HDPE or HDPE/LDPE/HDPE or HDPE/LDPE/ EAA.

The flexible pouches can be fabricated as flat pouch or pillow pouch or three side sealed pouch, or four side sealed pouch or stand-up pouch.

Semi Rigid Packs

Lined folding cartons having suitable laminated packed product will form a semi rigid package. Laminated pouch will provide the adequate protection to the product and outer carton will provide protection, strength and graphics and display for the primary package. Instead these laminates can be used as a lining material for the carton to get almost the same purpose as explained early. Combinations of paper/ met or PET/Poly or Paper/foil/ poly etc can be used for laminates. Tetra brik pack and 5kg carry home "Bag- in- Box" with a spout are available

Rigid Containers

These are the most common packaging forms used for packaging oils and vanaspathi. These are either made from tin as cans or rigid moulded plastic containers. Pigmented HDPE containers are suited to pack 1 to 2 kg and 5kg unit packs. Similarly blow moulded polyethylene terephthalate (PET) bottles are used for 1 to 2 kg oil unit packs. They have excellent clarity, superior impact resistance and low gas permeation. Moulded polyvinyl chloride (PVC) containers are also available for packaging oil. However, there is likely to have monomer migration problem in PVC containers when packed with oil.

Packaging materials most suited for specific products are given below:

1. Refined bleached and deodorized palm oil–multy layer film structures like 90 to 100 micron LLD or PP or HD with BA/PA/BA/EAA or 90 to 100 micron HD/LD/EAA (yellow pigment) or 90 micron LLD/HD/LD/HD/EAA (Green pigmented)

2. Refined ground nut oil, cotton seed oil, sun flower oil–PET/HD-LDPE or MET-PET/HD-LDPE

The above oil can be packed as 5kg unit pack in "bag-in-box" containing a flexible bag based on polyamide and PE as the bag placed in a chip board carton as the box to provide protection for the inner bag.

Spirally wound paper board composite containers lined with aluminium foil and polyethylene has found to be a rigid fancy pack for the refined oils.

Similarly polyester bottles (transparent) were found to have better protection than PVC or HDPE bottles with respect to quality and shelf life.

Of late Stand-up pouches with screw cap spout with handle forms an excellent convenient package for oils for its convenience and display property.

References

Baldev Raj, A.R. Indiramma and Balasubramanyam, N. (1986). Recent developments in packaging of oils in flexible consumer packages. Proceedings of the Symposium on Recent Developments in Food Packaging. AFSTI, CFTRI Mysore. 96: 102.

Indiramma, A.R. (1995) Packaging of edible oils and fat in the book Profile on food packaging. Published from CFTRI, Mysore.

Kumar K.R.(1992) Design and development of consumer packages for edible oil. Food Digest (15) 328.

Kumar K.R. Nagappan A. and Baldev Raj (1989) Evaluation of flexible packages to contain mustard oil. J. Food Science and Technology 26 (2)59.

V. Packaging Materials for Bakery and Confectionary Products

The basic requisites of packaging materials for biscuits and confectionary are:

☆ To give better shelf life to the product

☆ To prevent breakage and pilferage in transit

- ☆ To prevent contamination
- ☆ To enable easy handling
- ☆ To prevent moisture ingress
- ☆ To prevent the tainting of the product
- ☆ Material should be compatible with the product
- ☆ Material should be machinable
- ☆ Must be economical.

Main Types of Biscuit Packages

Envelope Type: An old mode of package which is almost obsolete now

Horizontal Flow Type: Most common mode of package

Package with Tray Inside: A package provided with rigid tray inside which will protect the product from breakage and spoilage. But it adds more cost to the product.

Pack made on vertical form, fill and seal machine.

Fancy Tin Packs: It is an attractive package and is sold as gift packages during festive seasons.

PET Jars: It is exclusively intended for small biscuits

Methods of Packing

- ☆ Manual packing
- ☆ Machine packing
- ☆ Nitrogen flushed packing

Machine packing is mostly carried out in an FFS machine; the sealing jaw of the FFS machine applies pressure, heat and cutting action. The temperature required for sealing depends on packaging material, dwell time between jaws and the pressure exerted by the jaws.

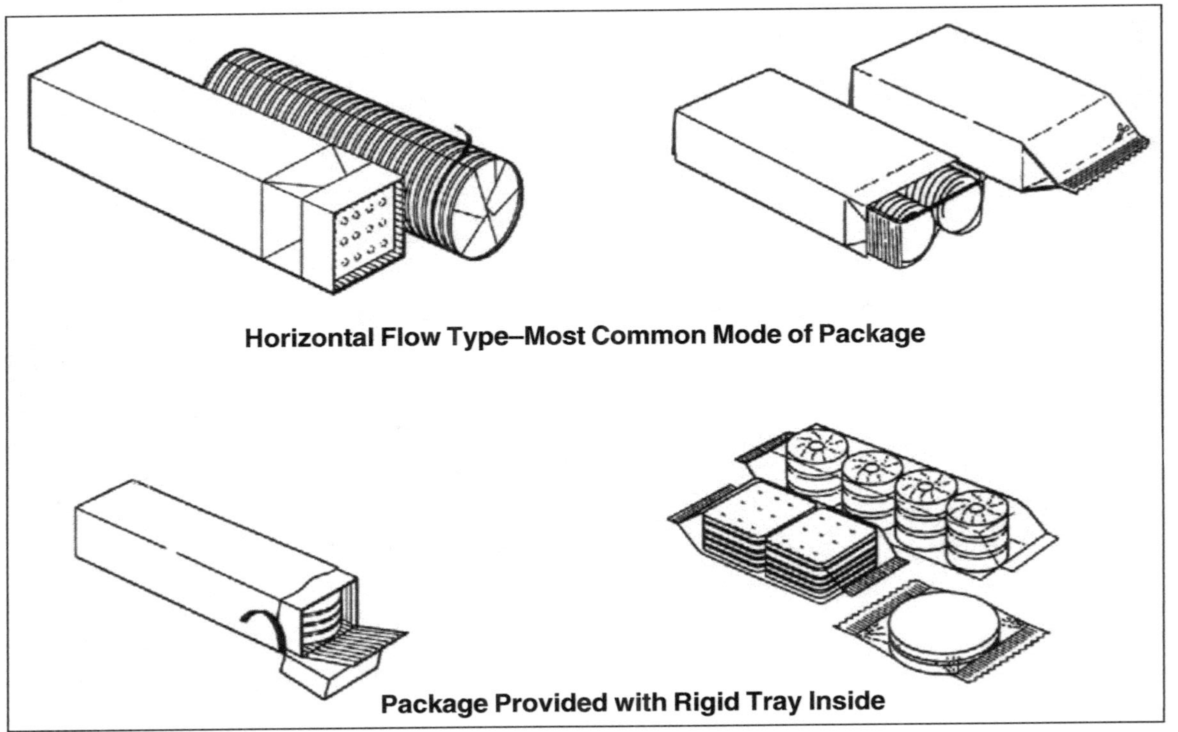

Horizontal Flow Type—Most Common Mode of Package
Package Provided with Rigid Tray Inside

Types of Materials Used in Biscuit Wrapping

1. Poly bags made from HDPE, LDPE and HM-HDPE mainly used for cheap biscuits and for manual packing.

2. Single layer of BOPP- used mainly as inner wrap for multi piece packs

3. BOPP/CPP laminate used extensively for glucose type biscuits.

4. BOPP/BOPP laminate it is used for a large variety of biscuits.

5. BOPP/metallised BOPP laminate gives better shelf life than BOPP/BOPP laminate, but it is more costly

6. BOPP/PET/metallised PET or metallised BOPP/POLY laminate will give excellent shelf life and excellent graphics on the metal back ground, but it is very expensive.

For packaging highly fragile small biscuits it can be done in nitrogen flushed bags made of vertical flow pack machines or packing directly in PET jars and the bags for nitrogen flushing can be made from BOPP or PET/metallised PET or BOPP/POLY or CPP laminate.

Cakes

These are packed mostly in BOPP/BOPP laminates. In certain cases, trays or poly coated boards are also used to prevent damage during transit.

Bread and Bun

Wrapping of Bread and Bun should ensure that the product is hygienic and free from dust and other contaminating agents and the wrapping should also be able to conserve the moisture content and prevent too rapid desiccation and staling and the material should be soft enough to run on wrapping machine at the same time, should provide physical protection against deformation of bread and bun. The material must be economical in cost.

Early packages used were simple news paper or poster paper which was cheap but encountered excess moisture loss; this was replaced by waxed sulphite paper with brand printing and graphics. Newer packaging materials include LDPE bags or sleeves which were slightly expensive but were really convenient and therefore all kinds of bread were packed. Poly propylene due to its transparency, rigidity and better machinability is found to be adequate and at thin gauges are economical. Coated BOPP and HM-HDPE films have great scope in future.

Confectionery

There are many types of confectionery available and each type needs different mode of packing. They can be classified as (1) Hard boiled sugar confectionery (2) toffees (3) chocolates (4) snack foods.

Packaging Requirements

The functional packaging material should protect against the following:

- ☆ Dust, dirt and other contaminants.
- ☆ Moisture pick up or loss resulting in sugar and fat bloom, stickiness, hardening and desiccation.
- ☆ Colour and flavour loss and tainting.
- ☆ Physical damages like dusting, scratching, chipping and breakage.

☆ Rancidity and infestation by insects.

☆ Migration of ingredients from the packaging materials to the product.

Hard Boiled Sweets

Hard boiled sweets are generally made almost entirely of sugar in an amorphous state. Maximum permitted moisture content of these sweet is 3 per cent therefore it has to be packed in moisture proof materials like waxed paper, MTTW cellophane, HM-HDPE, PP, or PVC films. Most suited wrappers for hard candy is twist wrap preferably from laminate of BOPP/met BOPP/ met PET, this has to be subjected to hot air blowing for proper twisting. Modern package can have an under strip made of waxed paper strip or metallised PET or metallised BOPP in contact with the product which is then over wrapped with the above laminate. For roll packs eg like Poppins/polo, two wrappers are used. The outer wrapper in the form of sleeves is closed with adhesives and the inner wrap is just folded. For outer wrapper foil/paper or paper/foil laminates are used and for the inner wrapper waxed paper or waxed paper/foil/poly laminates are used. For lollipop type products waxed paper can be twist wrapped over the sweet portion allowing the handle to protrude or by heat sealing two webs of a film around the product allowing the stick to protrude through the small opening. MST and MXXT cellophanes provide water vapour and aroma protection.

Medium Hard Sweets

These are low boiling or chewy candies such as toffees and caramels which are a mixture of sucrose and glucose added mostly with milk. For medium hard sweets the moisture content is between 8 and 15 per cent, so they require lesser protection than hard candies. Flexible grade of moisture proof cellophane is well suited. Waxed paper can also be used and the wrap can be either twisted or tuck folded.

Medium Soft and Soft Sweets

Gums and pastry products fall in this category. These products are having a moisture content ranging from 15 to 22 per cent. These products are therefore packed in bleached or opaque glassine paper which is having good grease barrier but a poor water vapour barrier property.

Chewing Gum

The outer sugar coating which is relatively stable can lose gloss and crispness due to absorption of moisture or the product can become hard and brittle due to the loss of moisture. Usually gums are packed in units of 2or 3 in waxed paper or in laminates of Paper/PE or Cellophane/ PE. Stick type chewing gums are individually wrapped in paper/ foil/lacquered wrap which prevent the stick from moisture absorption and aroma loss.

Chocolates

Chocolates when exposed to high humidities and temperature will cause condensation resulting in sugar bloom and softening of the product. Chocolates are preferred when the surface is glossy and smooth. Hence the packaging material should be sufficiently smooth and soft to minimize surface aberration and to provide good physical protection to the product. Similarly, since chocolate contains lot of fat there are chances to get easily tainted with foreign odour.

Small chocolates and chocolate covered nuts and sweets can be individually twist wrapped in soft cellophane and then filled in small bulk quantities in pouches of cellophane or polyolefin films.

Bars of chocolates are mainly wrapped in waxed paper, lacquered glassine or cellophanes or for better protection and shelf life laminates of paper/ foil/ lacquered wrap is the ideal one. Currently metallised polyester or nylon film in combination with PE is extensively used to wrap chocolates. Two web film package with the facing side containing transparent polyester/ PE and the reverse side having metallised PET/PE would provide the necessary appeal and protection. Small individual confections can be wrapped in waxed paper or cellophanes or polyester films and then contained in thermoformed trays with peel off lids which would provide the chemical and mechanical protection.

Chips

Potato, banana, jackfruit and other chips are low moisture snack foods and hence are very sensitive to moisture pick up. Hence, they require good waterproof and oxygen impermeable, odour resistant, grease resistant, materials having excellent physical strength to withstand against breakage, crushing and crumpling.

Potato and banana chips are generally packed in waxed glassine paper, coated cellophanes and high density polyethylene.

More functional flexible packaging materials include Cast PP and coated heat sealable BOPP, glassine/PE, cellophane/PE and both plain and metallised polyester/ PE and co-extruded films. For long shelf life laminated pouch of paper/foil/PE or cellophane/foil/ PE with or with out inert gas flushing are better.

Newer multiweb structures such as HD/LD/HD are well suited for Indian traditional snacks. Roasted, salted and spiced cashew kernels can be best packed in paper/ foil/PE or in PET/foil/PE laminated pouches which can offer 180 days life where as in cans a life of 10 to 12 months will be obtained.

Sweetmeats

Indian sweetmeats like sohan pari, sohan halwa, pedha are high in fat content hence requires good oxygen barrier and grease resistant materials. Generally these materials have only short shelf life due to development of discolouration, development of rancidity and fat bloom.

These are packed in PET/PE laminate if it is meant for 30 days, HDPE/foil laminate for 100days and in rigid cans for 225 days.

References

Anon (1976) Packaging of biscuits and confectionery. Packaging India Vol. 8(3) 13-16.

Anon (1979) Snack foods packaging a balanced approach. Snack Foods Vol. 50 (4) 588.

Anon (1983) Packaging of biscuits, bread and confectionery. Packaging India Vol. 15 (4) 31-34.

Kumar, K.R. and Balasubramanyam,N.(1979) Packaging profile for biscuits. Indian miller Vol.9(1) 7-12.

Kumar, K.R. (1995) Packaging of Bakery products, Confectionery and snack foods in the book Profile on Food Packaging. Published from CFTRI, Mysore.

VI. Packaging of Meat, Fish and Seafoods

Fresh meat is highly perishable and biologically active and a complex food for packaging. Its typical colour, odour flavour and texture are the important quality attributes which are needed to be preserved till it reaches to the consumer. Exposure to environmental oxygen will result in the formation of oxymyoglobin which has a bright red colour. Atmospheric oxygen is sufficient to cause deterioration in fresh meat colour within 3-4 days. Loss of moisture leads to undesirable dark reddish brown colour on the surface, loss in weight causes shriveling which will affect texture and juiciness in meat. Natural flavour of fresh meat has to be preserved. Pick up of undesirable flavour and odour due to the development of rancidity during storage and tainting from out side.

Therefore the packaging material should prevent moisture loss, must prevent bacterial contamination and pick up of foreign odour.

Materials and Techniques

Tray with Overwrap

Very common package is the plastic trays over wrapped with transparent film. The trays can be made from polystyrene foams. Use of blotters underneath eliminates the excessive juice accumulation.

Shrink Film Overwrap

Shrink films are used for wrapping larger and uneven

Prawn in Tray with Overwrap and Blotters Underneath

Shrink Wrapped Meat in Tray

cuts. Heat shrinkable PVC or PP, PVDC are used with the help of hot air tunnels to effect tight pack.

Vacuum

Vacuum packaging is recommended only when the P^H is below 5.9 otherwise hydrogen sulphide producing bacteria dominate and result in greening after 20 days. It is mostly

Dressed Chicken in Vacuum Pack

recommended for beef in developed countries. Materials most suited for vacuum packing are laminates of cellophane/poly, polyester/poly foil laminates and PVDC co- polymer.

Controlled Atmosphere Packaging of Meat (CAP)

In this package air is replaced by gases usually O_2, CO_2, or N_2 alone or in combination at controlled concentrations. Beef needs a high oxygen content to maintain bright red colour where as chicken needs little oxygen and reduced amount of carbon di oxide. Pork with its high fat content need less oxygen and some nitrogen for glow.

70 per cent oxygen, 20 per cent carbon di oxide and 10 per cent nitrogen gas mixture in a pack at 0°C slows down bacterial growth so that microbially caused discolouration, sliminess and aroma changes are eliminated. Packaging in CO_2 or $CO_2 + N_2$ mixture has an application in wholesale storage, where meat can be held anaerobically until required for display.

Fresh Sausage

The package requirement is similar to fresh meat. They are also subjected to oxidative colour changes, bacterial spoilage, dehydration and sensitive to light. Heat sealable cellophane, polyethylene film in the perforated form has been used to some extent for the pre packaging of the product. Now they are mostly sold out in natural casings *i.e.,* dried and cured intestine of the bovines. Gelatin incorporated cellulose casing and alginate formulations are also used.

Frozen Meat

This product is prone to dehydration leading to freezer burn and loss of surface texture. Therefore requires high moisture barrier material, the package must again should have barrier proofness against oxygen and light. During freezing the package should withstand the expansion and contraction during freezing and thawing. Shrink polyethylene and co- polymer of PVC and PVDC are usually used, nylon laminated to polyethylene can also be used.

Cured Meat

Coloured or opaque polyethylene, PVC and trays with shrinkable over wrap are used. For cooked meat and bacon high gas and water vapour barriers are required. For vacuum packing laminates of PE/PVDC/PA or metallised PET/Poly and PET/foil/poly are used.

Poultry

Just like any other meat this will also undergo deterioration like desiccation due to moisture loss, oxidative rancidity and microbial spoilage. Bags made of polyethylene and twisted or clipped are used for the dressed chicken. Trays made of cellulose acetate over wrapped with co polymer film of PVC and PVDC are used. Stretch films of PVC can be used to produce a package resembling shrink package. Packaging of large whole frozen poultry is usually done in PVC-PVDC copolymer film having a very high puncture resistance and ability to heat shrink by immersion in hot water. Shrink wrap will help to eliminate freezer burn and also minimizes oxidative changes. Frozen whole turkey is often sold in vacuum packaged shrink plastic film bags. Product is filled into the bag evacuated and immersed in hot water (90.5°C) to get skin tight wrap with no air pockets.

Packaging of Fish and Seafoods

Fresh fish are one of the most perishable commodity therefore must be refrigerated until consumed. A suitable fresh package must be:

1. Reduce fat oxidation and dehydration
2. Must provide less bacterial and chemical spoilage
3. Eliminate drip and prevent odour permeation.

The highly unsaturated nature of fish oils and fats makes them very susceptible to rancidification. Excess moisture loss leads to texture, flavour and spoilage. The characteristic fishy odour of trimethylamine is not present in fresh fish and is caused by enzyme action.

Rigid corrugated carton having wet strength has been replaced by wooden boxes for bulk shipment. Expanded polystyrene cartons (Thermo cole) are now used as a bulk transportation container as it is both light in weight as well having excellent insulation and hence can be used along with ice.

Retail Packaging

The package generally used is a polyethylene, wax or hot melt coating inside the carton gives high gloss on the product surface and the carton can be sealed. Frozen fish can be found in over wrapped trays. The trays can be made from moulded pulp or foam polystyrene or clear polystyrene. Individual fillets are over wrapped with cellophane or PVC film of 50 micron thickness.

Processed fish like salmon, tuna and sardines are usually canned. There are cans of various shapes like flat, oblong or oval and round shapes.

Smoked fishes are packed in nylon/poly or polyester/poly laminated pouches and sold or it can be placed in a boil-in-packs and evacuated and sold as frozen forms.

Modified Atmosphere Packaging of Fish

20-100 per cent carbon-dioxide atmosphere was found to be effective in controlling microbial growth of different types of fish. Packaging materials generally used are flexible films of nylon/ionomers, plastic moulded trays and over wrap with 75gauge PVC stretch film, polyethylene/poly laminate etc In certain places cold fillets are packed in a plastic moulded tray of 400 micron PVC laminated with 100 micron polyethylene film after flushing with 50 per cent carbon-dioxide and 50 per cent oxygen in a vacuum sealing machine.

Generally shrimps are weighed and filled into a unit carton which is lined with polyethylene, closed and frozen in a contact freezer and these are then packed in master carton of 5 ply waxed CFB box.

The development of individually quick- frozen shrimp (IQF) has expanded sales at retail levels in consumer packs as ready- to- eat products in Western and Scandinavian markets. So they are packed in deep drawn plastic pouches made of high thicker PVC materials on roll- fed machines with deep drawing of the bottom web are used or packed in printed preformed pouches made of laminate of oriented polyamide/poly material. Majority of these packs are of 75 to 200 gram vacuum packed IQF product and these pouches are then placed inside printed cartons made of bleached sulphite board which is plastic coated on both the sides.

References

Principles of Food Packaging, Editors Stanley Suchow and Roger C. Griffin (1980) The AVI publishing Company Inc., West Port.

Packaging of Foods (1986) Published by Indian Institute of Packaging, Bombay.

VII. Packaging of Fresh Fruits and Vegetables

Fresh produce properly treated, sorted, graded needs proper package for protection during transportation and

Traditional Packing House of Fresh Fruits and Vegetables

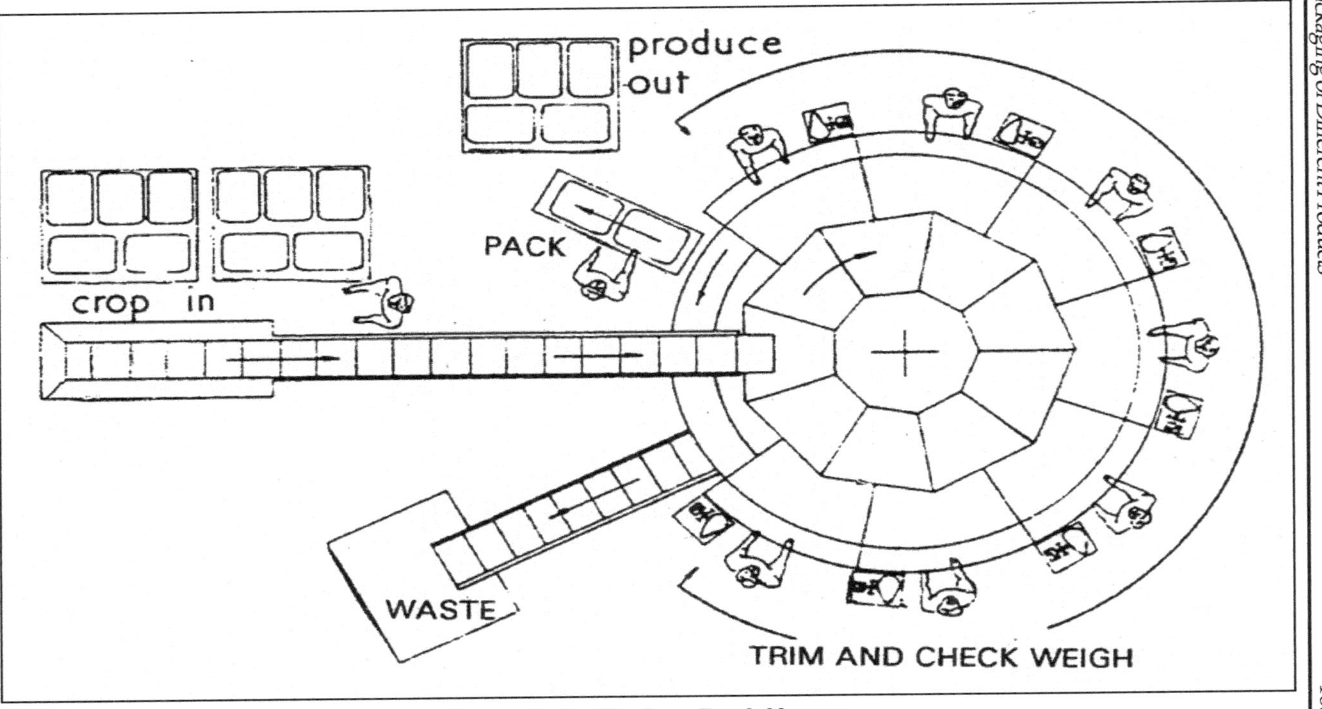

Layout of a Modern Pack House

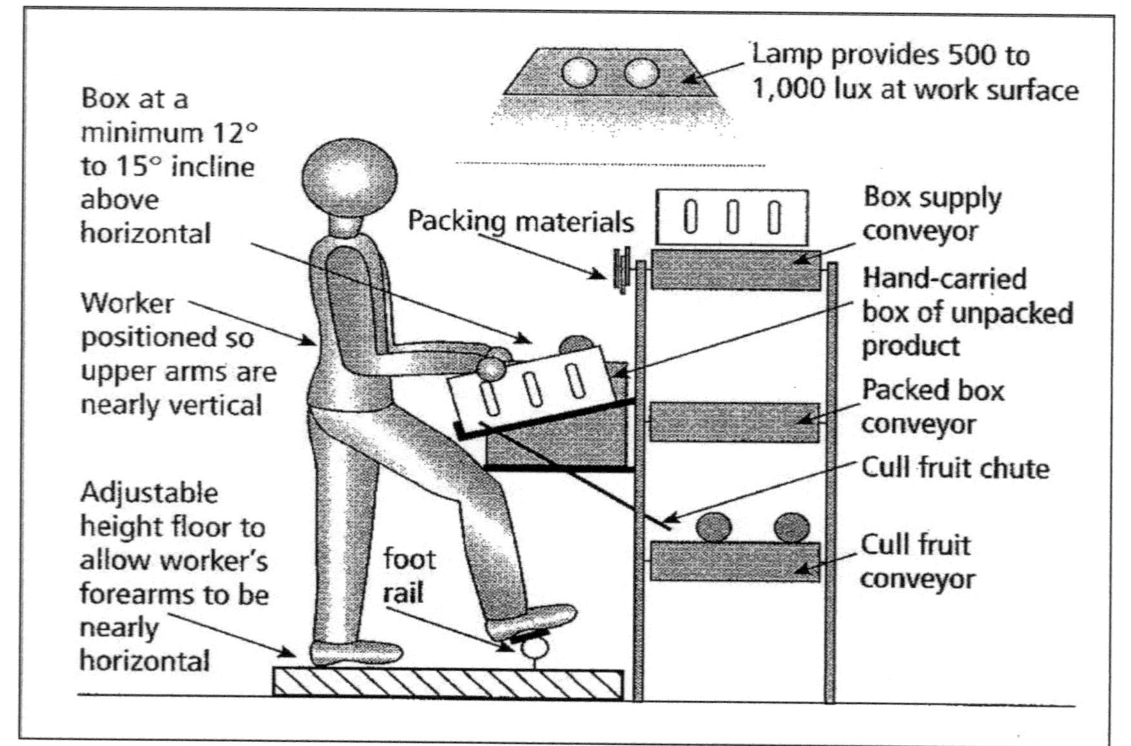
Lamp provides 500 to
1,000 lux at work surface
Box at a
minimum 12°
to 15° incline
above
horizontal
Packing materials
Box supply
conveyor
Hand-carried
box of unpacked
product
Worker
positioned so
upper arms are
nearly vertical
Packed box
conveyor
Cull fruit chute
Adjustable
height floor to
allow worker's
forearms to be
nearly
horizontal
foot
rail
Cull fruit
conveyor

storage. Methods of packaging can affect the stability of products in the container during shipping depending upon the extent of the container protecting the product. For examples, delicate and high priced products are packed in trays or in fibre board boxes, whereas other products are simply put directly in the boxes together.

Pre-packaging or consumer packaging is also convenient for retailers as well as customers and therefore adds value to the produce.

Pre-packaging generally provides additional protection for the products. At the same time, cover of non-biodegradable plastic trays and wrapping materials often seen in modern supermarkets, creates an extra burden of waste disposal and damages the environment.

Here you can observe that the pack house is not in an organized manner, packaging containers of all types along with the materials for packaging are accommodated in a small and congested room and the personnel are also sitting along with the material; where as in a modern pack house all the operations are well organized and stream lined without any cross over or over lapping.

Improved Packing Practices

- ☆ Pack securely to immobilize produce
- ☆ Do not overfill or under-fill packages.
- ☆ Avoid wood, nails, and staples due to possible food safety problems.
- ☆ Do not use very large packages
- ☆ Use shallow packages for delicate produce such as berries, grapes, summer squash and ripe stone fruits (single layer if possible).

☆ Do not overload packages or allow produce to bulge up over the top edges of the container.

☆ Consider using packaging materials such as trays, cups, wraps, liners and pads to help immobilize and protect produce, but don't overdo it.

☆ Do not block all ventilation holes of packages with fillers or liners, as this will interfere with cooling.

☆ Small, rigid plastic containers such as clamshell baskets (consumer sized containers placed into larger boxes) are useful for protecting berries, mushrooms and other delicate produce from damage and water loss.

☆ Choose packages carefully to avoid interference with cooling.

☆ Make sure the packages you use for garlic and onions have plenty of ventilation (mesh bags will provide the lower humidity environment required for proper handling).

☆ Packages for carrots benefit from using a plastic liner to reduce water loss.

☆ Label containers with your logo or farm name for enhanced marketing (have cartons printed or use inexpensive paper labels or stickers).

☆ Consider using packages that can be used as to display produce during marketing (printed sides, display materials on flaps).

☆ Packages should be vented, about 5 per cent of the surface area per side. Excessive venting, such as that found in some plastic crates, can lead to faster water loss.

☆ Consider the use of package liners to decrease venting in plastic crates to 5 per cent.

The Key Functions of Packaging

1. To assemble the produce in convenient units for handling

2. To protect the produce during handling, transportation, storage and marketing

Universally Standardized Requirements for Packaging

The package must have sufficient mechanical strength to protect the content during handling, transportation and stacking one over the other.

The package must meet handling and marketing requirements in terms of size, shape and weight in accordance with International Standards. The current trend is to avoid too many sizes and shapes of packages. Palletisation and mechanical handling makes standardization essential for economic operation.

☆ The material of the package must not contain any toxic chemicals which could transfer or produce toxin in to humans.

☆ The package should allow rapid cooling of the contents

☆ The package should be stable to moisture and high humidity.

☆ The packages should be stackable and interlock able.

☆ The packages should be re-usable or recyclable for easy disposability.

☆ The package should provide adequate ventilation

☆ Package should be of suitable standard depending on the market demand.

☆ It should be cost effective in relation to market value of the commodity.

Advantages of Packaging, Fresh Fruits and Vegetables

Packaging provides a beneficial modified micro-environment that helps in:

☆ Minimizing postharvest loss by: protecting against mechanical damage, microbes, pests, dust and air pollution, moisture loss, pilferage etc.

☆ Efficient handling and marketing

☆ Giving better appeal so as to promote sale.

☆ Providing hygienic condition within the package.

☆ Enhancing marketable distance and time.

☆ Protecting nutritive quality

☆ Preventing contamination by other commodities

☆ Providing information about the contents

☆ Ready to use facility

Classification of Packaging Materials

Traditional Ones

1. *Natural materials*: Bamboo, straw, palm leaves

2. *Natural and synthetic fibre*: Sacks, jute, cotton, woven plastic and paper.

3. *Wood*: Wooden crates, wire bound veneer and crates

Recent Ones

1. Corrugated fibre board: ventilated and non ventilated

2. Plastic crates

3. Molded trays, paper pulp and plastic

4. Net/Mesh bags/Sleeve pack

5. Plastic films/bags/boxes

Specialized ones

1. Cling film

2. Shrink wrap film/Stretch film

3. Flexible packaging materials for modified atmosphere packaging: Low density polyethylene (LDPE), Cellophane, Rubber hydrochloride, Polyvinyl chloride (PVC)

4. Fancy packaging materials.

Natural Materials

This type of packages includes baskets and other traditional containers made from bamboo, palm leaves, straw etc.

The characteristic features of such containers are:

Labour cost and raw material cost involved in making such containers are low and they provide good ventilation

Bamboo Baskets

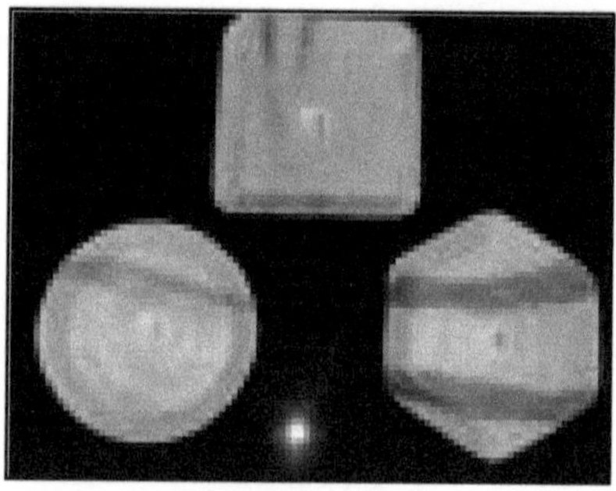

Areca Nut Leaves Sheath Trays

Drawbacks

☆ Difficult to clean if contaminated with decaying organisms

☆ Lack rigidity and bend out of shape when stacked

☆ Load badly because of shape

☆ Causes pressure damage when tightly filled

☆ Sharp edges or splinters cause cut and puncture damage to the contents.

☆ Less life, so should be replaced frequently.

Natural/Synthetic Fibres

Sacks

These types of packaging materials are made from jute or cotton or woven plastic or paper.

These are inexpensive and readily available, moreover they are reusable and have good load bearing capacity.

Jute Bags

Drawbacks

☆ Lack rigidity and handling can damage contents

☆ Too large for careful handling; if dropped it bruises the contents.

☆ Impair ventilation if stacked

☆ Difficult to stack on pallets

With these drawbacks also these are widely being used in Kerala for bulk packaging and transportation of fruits and vegetables. So a modification suggested is to place a woven bamboo about one meter.

Wood

Crates and wire bound veneers

These are the advantages in terms of characters like:

1. Rigid and reusable
2. Facilitates ventilation
3. Stack well on trucks if made to a standard size.

Wooden Boxes

Wire Bound Veneers

Drawbacks

1. Difficult to clean adequately for multiple use
2. Heavy and costly to transport
3. Sharp edges, splinters and protruding nails damage contents

Corrugated Fibre Board (CFB) Boxes/Cartons/Cases

Also known as cardboard/fibre board/pasteboard. It is made from a layer of corrugated fibre board sandwiched between 2 additional layers of fibre board.

The characteristic features of this most widely used and acceptable packaging materials that made it popular are:

Corrugated Fibre Board Box

Waxed Chip Board Cartons

☆ Lightweight (20 to 25 percent) lighter when compared to wooden box of similar size).

☆ Excellent cushioning property and smooth nonabrasive surface.

☆ Low cost, reusable and recyclable

☆ Excellent printability

☆ Easy to set up for storage

☆ Available in wide range of sizes, designs and strengths.

Drawbacks

☆ Easily damaged by careless handling and stacking

☆ Weakens on exposure to moisture

Plastic Crates

These are moulded from high density polyethylene (HDPE). This packaging material has the following features:

☆ Reusable, strong, rigid, smooth

☆ Easy to clean

☆ Good ventilation

☆ Can be made to stack when filled and nest when empty, so space saving.

Plastic Crates

Drawbacks

☆ Costly

☆ Mostly imported adding to the cost

☆ Not foldable

☆ Deteriorate when exposed to sunlight unless treated with UV inhibitor.

☆ Though costly, capacity for reuse makes it an economical investment.

Moulded Trays

It may be moulded from paper pulp or plastics. It is suitable for packaging individual fruits and vegetables.

A. Paper Pulp Moulded Tray

☆ Made from recycled paper with starch binder

☆ Inexpensive

☆ Absorbs surface moisture from the product so it is good for small fruits and berries which are easily damaged by water.

☆ Biodegradable and recyclable

Moulded Trays with Apple

Thermoformed Plastic Trays

Advantages

☆ Rigid packaging

☆ Immobilization of the produce within the pack

☆ Suitable for microwave cooking without grease and loss of vitamins.

Thermoformed Plastic Trays with Fruits

Method of Packaging

The produce is filled in trays, over wrapped with a heat shrinkable film and passed through a hot tunnel.

The most commonly used micro-oven able packages are crystallized polyester (CPET) in the form of trays. They are flexible in shape and design and resistant to oil and greases. Polypropylene co-extruded with barrier resins like Ethyl Vinyl Alcohol (EVA) is used when a longer shelf life is required.

Plastic Film Bags

Polyethylene and polypropylene bags are mainly used, which have the following characteristic features:

- ☆ Low cost
- ☆ Widely used for consumer size packs in fruits and vegetable marketing
- ☆ Retain water vapor so as to reduce water loss from the contents.

Drawbacks

- ☆ No protection from injury caused by careless handling

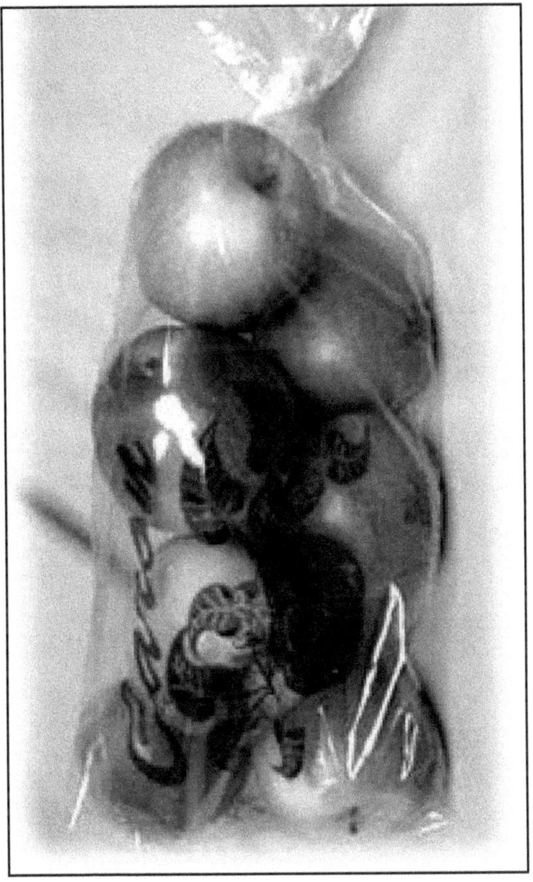

Plastic Bag containing Apples

☆ Heavy buildup of condensation may lead to decay

☆ Rapid build up of heat if exposed to sunlight

☆ Permit only slow gas exchange. Vapor and heat along consumer packs of plastics are not recommended in tropics except in stores with refrigerated display cabinet.

Plastic Boxes

Ideal for consumer packs and has good sales appeal

Net/Mesh Bags

Widely used for packing fruits like apple, citrus, guava, sapota etc. and have the following features:

- ☆ Sturdy
- ☆ Low cost
- ☆ Uninhibited airflow
- ☆ Attractive display which stimulates purchase.

Plastic Box with Okra

Drawbacks

- ☆ Large bags do not palletize well whereas small ones do not efficiently fill the space inside CFB boxes.
- ☆ Do not offer protection from rough handling
- ☆ Little protection from heat and contaminants

Sleeve Packs

Heat shrinkable film of 1.5 to 2 mm thickness is used. The following are the features of this type of packaging.

☆ Immobilization of packed fruits

☆ Superior visibility that gives a good sales appeal

☆ Low cost

☆ Better protective qualities

Cling Film

Polyethylene film of 15 microns thickness is used. Features that make it an ideal packaging material are:

☆ Low water vapour transmission rate

☆ High gas permeability

☆ Intimate package with the individual produce

☆ Keeps the produce fresh, dust and insect free

☆ Self-sealing types are available commercially.

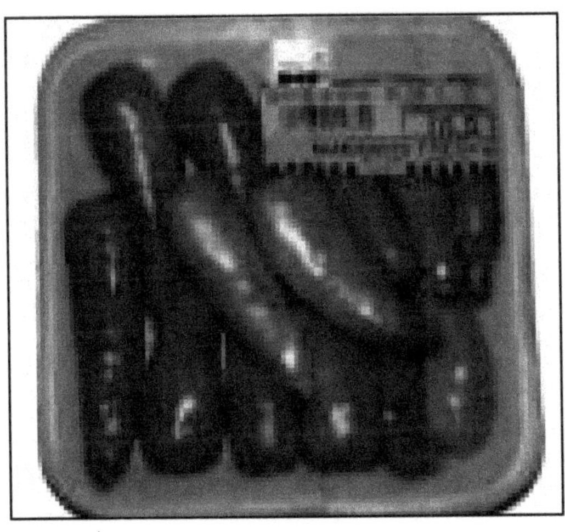

Carrot Individually Wrapped with Cling Film

Shrink Film or Stretch Film

Principle involved is plastic film like polyethylene, poly styrene, polyvinyl chloride, polyester, rubber

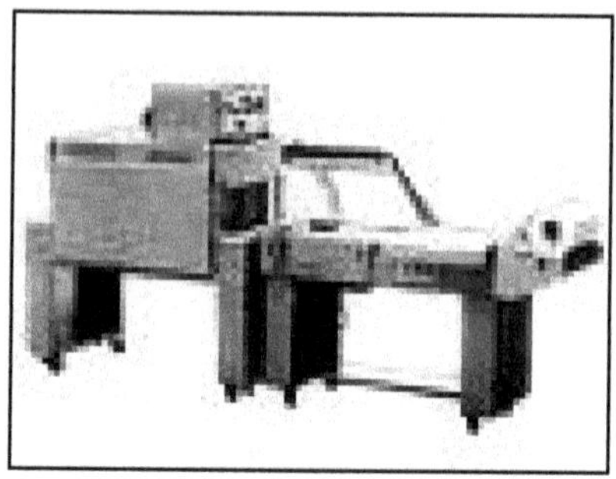

Shrink Wrapping Machine

hydrochloride etc have heat shrinkable nature. By stretching the film under controlled temperature and tension the film which is wrapped over the produce stretches and then contract by cooling.

The characteristic features are:

1. Can be used as over wraps on individual fruits and over consumer size trays or pack

2. Shrink on exposure to moderate temperature

3. Heat supplied by electric resistance coil is just enough to shrink the film but not enough to harm the produce.

4. Mineral impregnated polyethylene film adsorbs and removes ethylene gas and has excellent permeability and good deodorizing properties.

5. The addition of anti-fogging treatment to the film reduces the formation of water drops and the protection for mould and bacterial growth.

Benefits of Shrink Wrapping

Prolongs the life, freshness and colour of packaged fruits, vegetables and cut flowers.

Possible to store and ship ethylene generators and ethylene sensitive products in proximity to each other.

The film maintains high humidity levels reducing water loss from packaged produce.

The potential for mould and bacteria growth and spoilage is reduced by anti-fogging treatment.

- ☆ A good surface for stick on labels
- ☆ Protects the produce from diseases
- ☆ Reduce mechanical injury
- ☆ Shrink wrapped lettuce keeps good up to 6 weeks and carnation up to 3-4 weeks (these are classified as highly perishable commodities).

Active Packaging of Fresh Fruits and Vegetables

Active packaging is a group of technologies in which the package is active and is actively involved with food products or interacts with internal atmosphere to extend shelf life while maintaining quality and safety.

Active packaging, some times referred to as 'interactive' or 'smart' packaging, which is intended to sense internal or external environmental changes and to respond by changing its own properties or attribute.

Potential Technologies Used in Active Packaging

Oxygen Scavenging

Most important objective of active packaging has been the removal of oxygen using various techniques *viz.*, absorption, interception and scavenging. The most widely researched and patented area of active packaging is the use of oxygen absorbing systems.Mitsubishi Gas chemical company was the first to launch ferrous based oxygen scavenger sachets under the trade name 'Ageless' and till date they are regarded as the pioneer in the oxygen scavenging technology.

Antimicrobial Packaging

Active packaging system comprising incorporation of antimicrobial agents into polymer surface coatings and surface attachments can be of immense value. Most widely publicized antimicrobial agents are silver salts on zeolite incorporated in plastic films and sheets or on material surfaces and into absorbent pads for fresh meal and produce.

Ethylene Control

Ethylene is a hormone, which accelerates ripening in fruits followed by senescence. Removal of ethylene from plant environment can significantly retard post harvest catabolic activity in fresh produces and modified atmosphere preservation process. 'Frisspack' is a paper incorporated with $KMnO_4$ for use in corrugated fiberboard cases.Activated charcoal impregnated with a palladium catalyst and placed in proper sachets effectively removes ethylene by oxidation from packages of minimally processed kiwi, banana, broccoli and spinach.

Moisture Control

Since the metabolism of fats and carbohydrates, produces water, condensation or sweating is a problem in many kinds of packaged food, particularly fresh fruits and vegetables or minimally processed prepared foods. Use of humectants between two highly permeable layers of a plastic film has been found to buffer the humidity inside the food package.

In Package Fumigant

1. Slow release system developed by CFTRI consists of a mixture of 99g KMS and one gram citric acid. Approximately 500±5mg of mixture is packed in craft paper bag of 4″ x 2½″ and folded and stapled. Eight such unit packets (4 g of material) were used for 4 kg of grapes.

2. Combination of slow and quick release system contains an equal proportion of a mixture of 100 g KMS and 60g CA (Citric Acid) which releases SO_2 quickly. This gave significantly better protection to grapes without adversely affecting quality during transportation and controlled spoilage in conventional packaging.

Gas Permeability Control

O_2 Control

Package material designers have developed high oxygen permeable packaging material. O_2 permeation rates of 18.06 in m^2/ day and CO_2 permeation rates of 7.74 in m^2/day are achievable and effective at various controllable ratios for high respiration rate for fresh produces such as broccoli, mushrooms, asparagus and strawberries.

CO_2 Control

It is a complementary approach to O_2 control. A mixture of iron powder and $Ca(OH)_2$ in the form of sachet, which scavenges both O_2 and CO_2, has been used to pack fresh ground coffee in flexible bags.

Active packaging is bound to emerge as a concrete step in the evolution process of advanced packaging technologies to meet the requirements of modern innovative food processing.

Reference

Abe, K. and Watada, A. F. 1991. Ethylene absorbent to maintain quality of highly processed fruits and vegetables. *J. Fd Sci.* 56: 1589-1592.

Anand, J.C. and Maini, S. B. 1982. Fibreboard packaging for fruits. *Ind. Hort* 31-55.

Banks, N.H. 1984. Some effects of Tal Prolong coating on ripening of bananas. *J. Exptl. Botany* 35: 127-137.

Brody, A. L,, Strupinsky, E. R. and Kline, L. R. 2001. Active Packaging for Food Applications. Technomic Publishing Co., Inc. Lancaster, USA.

Floros. J. D., Dock, L. L. and Hall, J. H. 1997. Active Packaging Technologies and Applications. *Fd. Cosmetics and Drugs Packaging.* 20: 10-17.

Joshi, G.D. and Roy, S.K. 1984. Standardization of packaging of fresh mango fruits (cultivar Alphonso) for transportation and storage. *Proceedings of the National Conference on Packaging of Fresh and Processed Foods.* March 2 and 3 1984, Calcutta.

Maini, S.B Lal and Anand, J.C. 1993. Fruit packaging. Advances in Horticulture. (Ed.) Chadha, K.L. and Pareek, O.P., Malhrotra Publishing House, New Delhi. (4): 1775.

Peleg, K. 1985. Produce handling packaging and distribution. AVI Publishing Company, Connecticut. p 233-236.

Rooney M. L. 1995. Overview of Active Food Packaging. In Active Packaging (Ed. Rooney, M. L.). Blackie Academic and Professionals, Glasgow, UK,120-123 pp.

Satyan, S.H., Scott, K.J., Graham, D. 1992. Storage of banana bunches in sealed polyethylene tubes. *J. Hort. Sci.* 67: 283-287.

Shetty, K.K., Klowden, M.J., Jang, E.B., Koshan, W. 1989. Individual shrink wrapping technique for fruit fly disinfestation in tropical fruits. *Hort. Science* 24: 317-319.

Chapter 7
Packaging Laws and Regulations

Packaging laws and regulations affecting food products are mainly covered under:

1. The standards of weights and measures Act. 1976, and the standards of weights and measures (packaged commodities) rules, 1977 (SWMA)

2. The Prevention of Food Adulteration Act, 1954, and the Prevention of Food Adulteration Rules,1955 (PFA)

3. The Fruit Products Order, 1955(FPO)

4. The Meat Food Products Order, 1973 (MFPO)

5. The Agmark Rules

The packaging regulations under SWMA are applicable to all kinds of commodities including food stuffs; this regulation emphasis on regulating quantities to be packed and related declarations.

One of the most important rule under SWMA is that the commodities to be packed for retail should be packed in standard specific quantities as given under the rule for each commodity. However, the Central Government can authorize pre-packaging in quantities other than those specified on technical ground.

Class-A Commodities

(*a*) The net weight or volume of which does not exceed 25 g or 25 cc.

(*b*) The flow properties, density or both of which cannot be maintained constant except with the help of considerable special technical effort.

(*c*) Containing several substances of different densities or different physical phase in packaging of which requires several operations;

(*d*) Which require several operations for packaging;

(*e*) Which, after they have been packed,are subjected to additional processing such as heat treatment, which is likely to effect the weight of the commodities in irregular or unpredictable manner.

(*f*) Composed of pieces, fragments or grains in which the maximum weight of each piece, fragment or grain is greater than or equal to the maximum permissible error corresponding to the net quantity contained in that package, if the commodity,is considered as belonging to class-B.

(*g*) Liquid commodities.

Class-B Commodities

Any commodity which does not fall within class-A, for commodities declared by number, the maximum permissible error is 2 per cent of declared quantity.

Label Declarations

The SWMA requires certain declarations to be made on every retail package, which include common/generic name of the product, net quantity, retail sale price, unit sale price, month and year of manufacturing or pre-packing and name and address of the manufacturer or the packer.

Product Name

The common name or generic name of the product cannot be coined for fancy brand name and it should be intelligible to all.

Net Quantity

The net quantity should be expressed in standard units of weight, measure or number or a combination of weight, measure or number as would give as accurate and adequate information to the consumer with regard to the quantity of the commodity contained in the package. The declaration of the quantity should be in terms of:

1. Mass, if the commodity is solid, semi-solid, viscous or a mixture of solid and liquid.
2. Length, if the commodity is sold by linear measure,
3. Area, if the commodity is sold by area measure.
4. Volume, if the commodity is liquid or if it is sold by cubic measure.
5. Number, if the commodity is sold by number

Where ever it is necessary to communicate to the consumer any additional information about the commodity, such information should also appear on the same panel in which the other information, as required by the rules, have been indicated. Additional information like the following is necessary to be given:

1. In the case of a concentrate, the dilution ratio of that concentrate.

2. In the case of dehydrated commodity, the reconstitution ratio of that commodity.

3. In the case of a package containing, say gulab jamun mix the number of gulab jamuns that may be obtained from the mix and weight of each gulab jamun.

In declaring the net quantity, the weight of the wrappers and materials other than the commodity should be excluded. However, when the package contains a large number of small items of confectionery, each of which is separately wrapped, the net weight may include the weight of such immediate wrappers, if their total weight does not exceed:

1. 8 per cent in case of wrappers made of waxed paper or any other paper with wax or aluminium foil under strip.

2. 6 per cent in case of any other paper of the total net weight of all the items of confectionery contained in the package minus the weight of immediate wrapper.

The declarations of those commodities which are likely to undergo significant variations in weight or measure on account of environmental or other conditions may be qualified by the words "WHEN PACKED". The declaration of quantity should not contain any word or expression which tends to create an exaggerated misleading or inadequate impression as to the quantity of the commodity. For example, the expressions like "minimum", "not less than", "average". "family", "economy", "large", "extra", "king", or any other word or expression similar nature should not be used.

The declaration of month and year of manufacturing or pre packaging should be given for all food products except in case of returnable consumer products like liquid milk in glass bottles, ready-to-serve fruit beverages.

Retail Sale Price

Retail sale price is the maximum price at which the commodity in packaged form may be sold to the ultimate consumer and declaration should carry the words "Maximum Retail Price" (MRP)....inclusive of all taxes.

Unit Sale Price

Unit sale price, the price per specified unit of weight, measure or number, should be given if the packed quantity is not a standard package size. Also, the declaration is not required, in case of package of those commodities which are not specified in the third and sixth schedule related to the standard package sizes. Further it is not required in case of commodities not specified in the third schedule but are packed in quantities of 100g, 200g, 500g, 1 kg, 2kg, 5kg or in multiples of 5kg or in 100ml, 200ml,500ml, 1litre, 2litre, 5litre and in multiples of 5litre.

General Guidelines on giving Declarations

All the declarations required to be made under SWMA should appear on the principal display panel. The principal display panel is defined as that part of a label which is intended or which is likely to be displayed, presented or shown or examined by the consumer under the normal or customary conditions of display, sale or purchase of the commodity contained in the package. Every declaration which is required to be made on a package should be legible, prominent, definite, plain and unambiguous and should be given in a specified minimum size depending on the area of the principal display panel.

Prevention of Food Adulteration (PFA)

The PFA is basically intended to protect consumer's health and safety. Under these rules, an article of food is termed adulterated if it has been prepared, packed or kept under insanitary conditions whereby it has become contaminated or injurious to health, or if the container of the article is composed, whether wholly or in part, of any poisonous or deleterious substance which renders its content injurious to health. Accordingly, food packed in containers made of plastic materials not conforming to the following Indian Standards Specification, are deemed to render it unfit for human consumption.

IS: 101146-1982 specifies the quality of polyethylene in contact with food stuffs

IS; 101142-1981 specifies the quality of Styrene polymers in contact with food stuffs

IS: 10151-1982 specifies the quality of Polyvinyil chloride in contact with food stuffs

IS:10910-1984 specifies the quality of polypropylene in contact with food stuffs

IS: 11434-1985 Specifies the quality of ionomer resins in contact with food stuffs.

Product Name

Every package of food stuff should carry the name, trade name or description of food contained in the package. In case of proprietary foods which have not been standardized under PFA, the name of the food or category under which it falls as per these rules, should be mentioned.

Ingredients

The names of ingredients used in the product should be listed on the label in descending order of their composition. In case of artificial flavouring substances, the chemical names of the flavours need not be declared, but in case of natural flavouring substances or nature –identical flavouring substances, the common names of the flavours should be mentioned, ingredients' falling in certain classes such as edible oils, spices and condiments, permitted preservatives, flavours etc can be listed using their class titles. However, packages weighing 20g or less and liquid products marketed in bottles which are recycled for refilling are exempted from the requirement of ingredients declaration.

Name and Address

The name and complete address of the manufacturer or importer or vendor or packer should be declared on the pack of the product.

Net Quantity

The label of the pack should carry the declaration of the net weight or measure or number of the contents, except in the case of biscuits, breads, confectionery and sweets where they can be expressed in terms of either average weight or minimum net weight. In case of packages containing large number of small tems of confectionery, each of which is separately wrapped, PFA has similar provisions as those under SWMA. Also, in case of carbonated water containers and packages of biscuit, confectionery and sweets, containing more than 60g, but not more than 120g, particulars of net quantity need not be given as per rules of PFA.

Batch Number

Declaration of batch number or code number or lot number which is a mark of identification by which the food can be traced in manufacture and identified in distribution, is mandatory under PFA rules. However, this rule is exempted in case of carbonated water containers and packages of biscuits, confectionery and sweets, containing more than 60g and packages containing bread and milk including sterilized milk.

Month and Year

The month and year in which the commodity is manufactured or pre-packed should be given for all food products except in case of (1) any bottle or package containing liquid milk, liquid beverage having milk as ingredient, soft drink, ready- to –serve fruit beverages or the like which is returnable by the consumers for refilling except in case of sterilized or UHT milk (2) any package, containing bread or UHT milk (3)any package, containing bread and any un canned package of (a) vegetables,(b) fruit (c) ice-cream, (d) butter (e) cheese (f) fish, (g) meat or (h) any other like commodity (4) any package of food where the net weight or measure of the commodity is 20g or 20 ml or less.

Nutritional Food

When the food product is claimed to be enriched with nutrients such as minerals, proteins or vitamins, the quantities of such added nutrients should be mentioned.

Infant Food

Infant food under PFA rules means any food which may be used for partial or total replacement of breast milk, commonly called breast milk substitute and includes infant milk food, infant formulae and any food suitable as a compliment to breast milk, to meet the nutritional needs of the infant after four months of age, commonly called "complimentary food" "breast milk supplement "or "weaning food".

Special labeling rules for infant food on packages under PFA are:

1. Packages of breast milk substitutes should not have a picture of infant or picture of mother breast feeding infant or other forms of picture or text which may idealize the use of such product. The terms "Humanised"or "Maternalised" or similar expressions should not be used.

2. Packages of infant foods should not carry expressions like "Full Protein Foods" "Complete Food" or "Health Food"

3. All infant food packages should carry the declaration "BREAST MILK IS BEST FOR YOUR BABY". The colour of the statement should be different from that of background text. The statement must be prominent and conspicuous in central panel of the label. In case of weaning food, the statement should be prominent and conspicuous on any panel of the label, besides carrying an accompanying statement "From four months of age, give baby cereal food in addition to mother milk"

4. The label should contain instructions for appropriate and hygienic preparation including cleaning of utensils, bottles and warning against health hazards.

Other Labeling Rules under PFA

All declarations required under PFA Rules should be either in English or Hindi. There should be surrounding line enclosing a declaration and the surrounding line should not be less than 1.5 mm. The types used for declaration should be of such dimension that it is conspicuous to a reader and should not be in any case less than 3mm in height. The word "synthetic" when ever is used shall be of the same size as used for the name of the product. In general, a label should not contain any statement, claim, design, device, fancy name or abbreviation which is false or misleading in any manner, except in case of established trade or fancy names of confectionery, biscuits, aerated water.

The labels should not contain ant reference to PFA or statements suggesting that the food is recommended, prescribed or approved by medical practioners. The word "Imitation" or any word or words implying that the article is a substitute for any food, unless the use of such word is specifically permitted under these rules.

Any fruit juice, syrup, squash, beverages or cordial or crush which does not contain the prescribed amount of fruit juice, should be described as synthetic product, and consumer should not be misled that it is a fruit product. Neither the word "FRUIT" should be used in describing such product nor can it be sold under the cover of a label which carries picture of any fruit. Carbonated water containing no fruit juice or pulp should not have a label which leads the consumer into believing that it is fruit product. The word "pure" or any word or words of same significance should not be used on a package that contains an imitation of any food.

Apart from the above rules, PFA specifies certain declarations to be given specified formats for certain food products such as coffee-chicory mixture, condensed milk etc. Foods containing Monosodium Glutamate should be declared on its label that the food contains added Monosodium Glutamate and unfit for infants below 12 months.

Fruit Products Order (FPO)

The FPO is concerned with fruit and vegetable products including synthetic beverages, syrups, sharbats and vinegar. The objective of this law is mainly to regulate the quality and hygiene of these products. All the fruits and vegetable products when packed will have to have the labels approved by the authorities concerned, and carry the license number allotted. When a bottle is used as the package,it should be sealed that it cannot be opened without destroying the license number, and the special identification mark of the manufacturer should be displayed on the top or neck of the bottle. The batch/code number along with the date of manufacturing should also be declared.

As contained in PFA, FPO also prohibits use of any statement, design or device which is false or misleading concerning the fruit product. Synthetic products associated with fruits and vegetables should clearly be marked "SYNTHETIC" and this word when ever used should be bold and in the same size and colour of the letters used for the name of product, and should immediately precede such name.

Meat Food Products Order (MFPO)

This order is similar to FPO, regulates the licensing and labelling of all meat products. All labels have to get approved by the licensing authority, and the license number and category of manufacture should be declared on the label.

The name of the product, always a common name understood by the consumer, should be given along with net quantity. Trade names should have prior approval of the licensing authority. When any preservative or colouring agent is used, a statement to that effect should be given. When permitted artificial flavouring agent is used, the word "Artificially Flavoured", should appear on the label in prominent letters and in continuance of the name of the product. The list of ingredients should also be given. Terms which may bear some geographical significance with reference to a locality other than in which either in the factory is located or the product is manufactured, can be given on the label after being qualified by the word "STYLE", "BRAND"' or "TYPE" as the case may be. No statement, word, picture or design which may convey a false impression or give a false indication of origin or quality, can appear on the label.

Agmark Rules

Agmark rules are related to the quality specifications and needs of certain agricultural products to be eligible for Agmark certification. They also specify the type of packages that can be used for various products and labelling declarations that have to be given. Some of the food products that have been covered under these rules are edible nuts, ghee, honey, pulses, spices and condiments and vegetable oils.

References

G.C.P.Ranga Rao (1995) Packaging laws and regulations in the book "Profile on Food Packaging" Published from CFTRI, Mysore.

Index

**Leaf Based
(Page No. 16)**

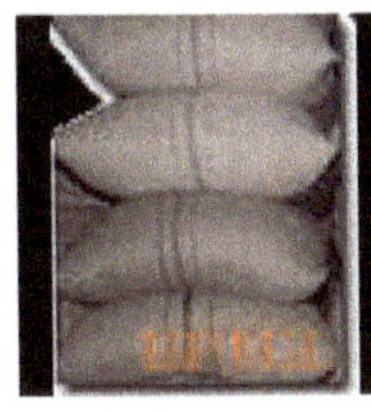

**Jute Based
(Page No. 16)**

**Wood Based
(Page No. 16)**

**Clay Based
(Page No. 16)**

**Shipping Container
(Page No. 19)**

**Wooden Shipping
Container
(Page No. 19)**

**Drum
(Page No. 21)**

**Barrel
(Page No. 21)**

**Wooden Pallets with
Modular Arrangements
(Page No. 25)**

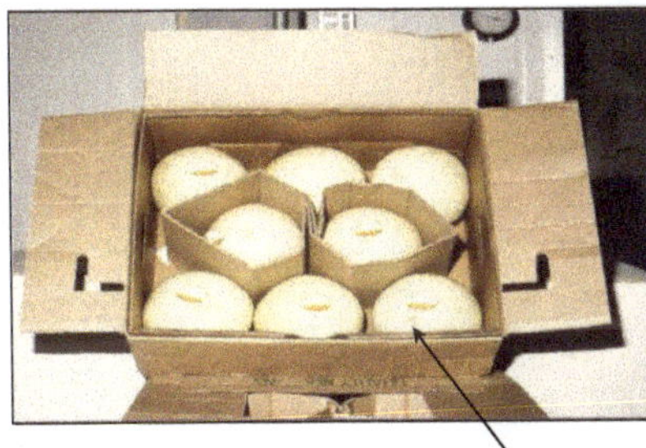

(Page No. 27)

**Hexagonal
Cells**

Spout of the inners flexible bag

Outer carton box

(Page No. 24)

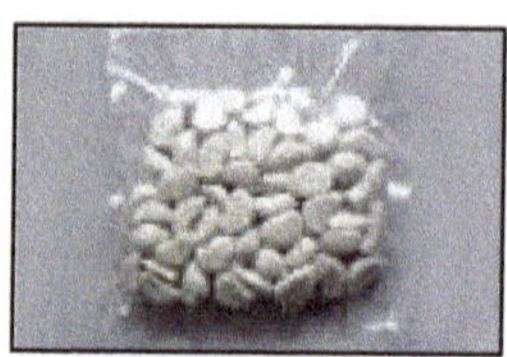

Vacuum
Packed Nuts
(Page No. 29)

Nitrogen Filled Gas
Pack for Fragile
Materials like
Potatochips
(Page No. 29)

Minimally Processed Vegetables in MAP
(Page No. 30)

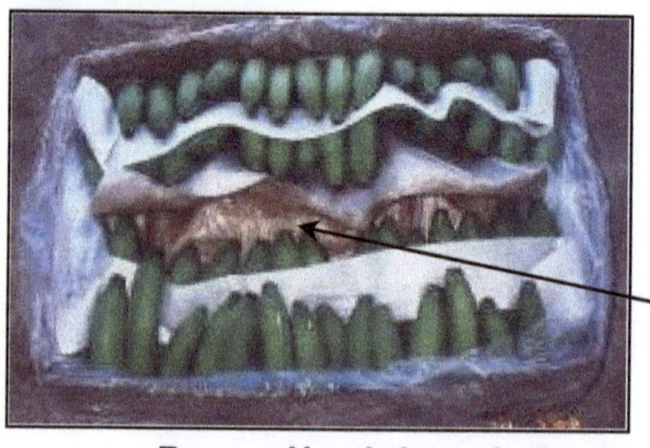

Sachet containing
ethylene scrubber

Banana Hands in an Active Package
(Page No. 31)

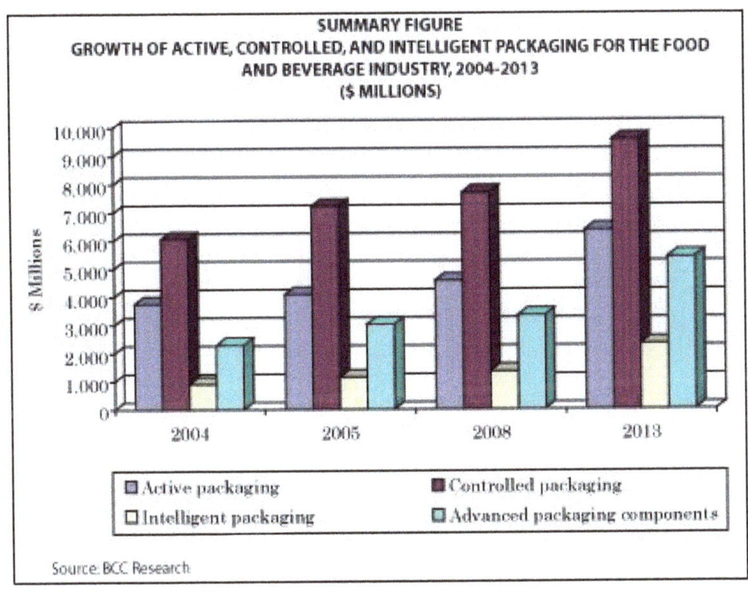

(Page No. 33)

Fruit Packed with "Ripesense" Label
Red: Unripe; Orange-red: Half-ripe; Yellow: Fully ripe
(Page No. 36)

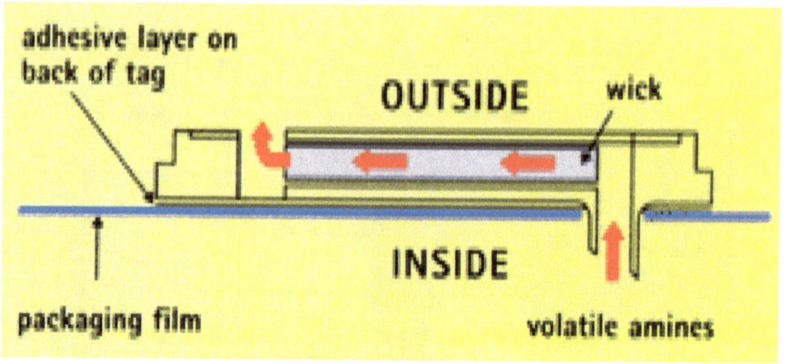

(Page No. 37)

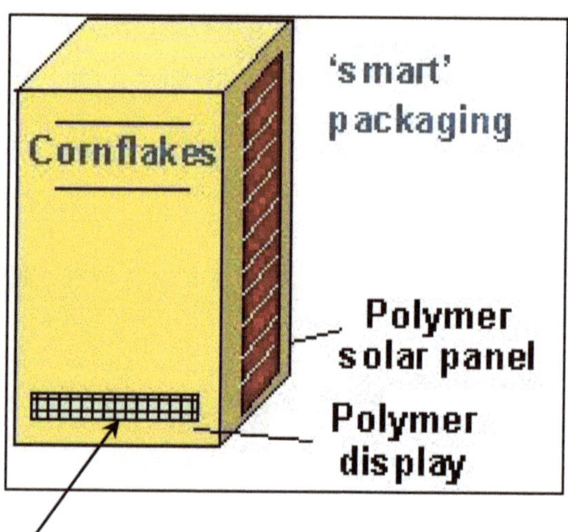

Polymeric Material which will Change Colour at Certain Stress
Threshold Levels indicating the Spoilage of the Product
(Page No. 39)

Aseptically Packed Fruit
Juice in Tetra Brick Pack
(Page No. 30)

Edible Chocolate Tray made
from Corn Starch
(Page No. 41)

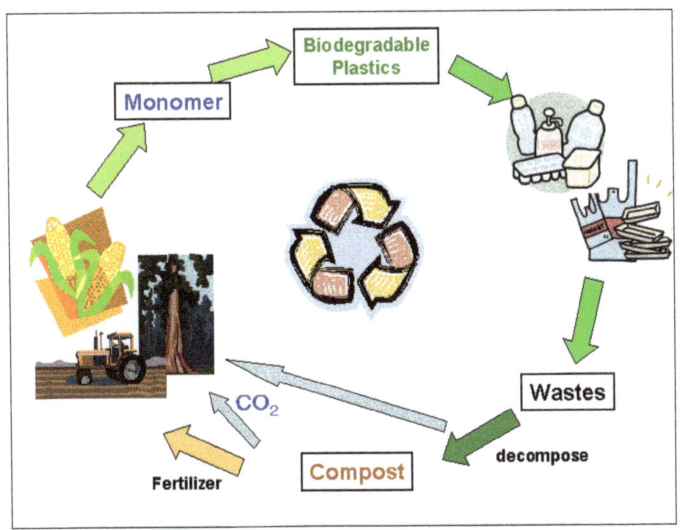

A Schematic Diagram Illustrating the Decomposition of a Biodegradable Plastic (Page No. 56)

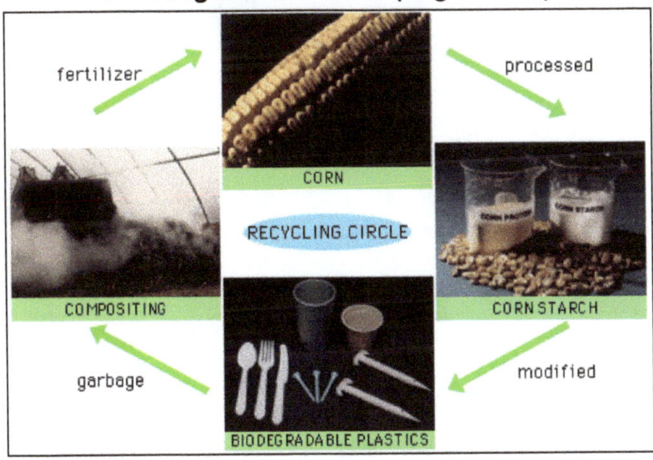

Recycling of Biodegradable Plastic (Page No. 56)

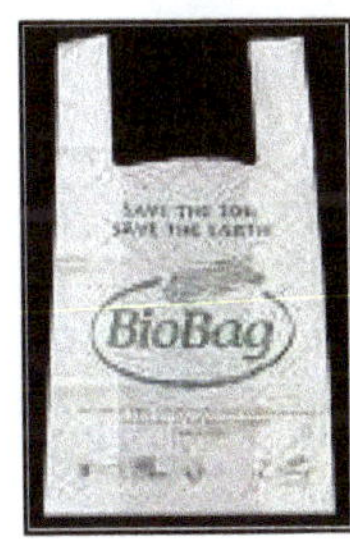

A Carry Bag of Biodegradable Material (Page No. 57)

Deep drawn aluminum cans with liquid products (Page No. 91)

Shallow drawn two piece can with easy open end (EOE) lid (Page No. 90)

Shallow Drawn Aluminum Can with Wide Mouth to Fill Solid Products (Page No. 90)

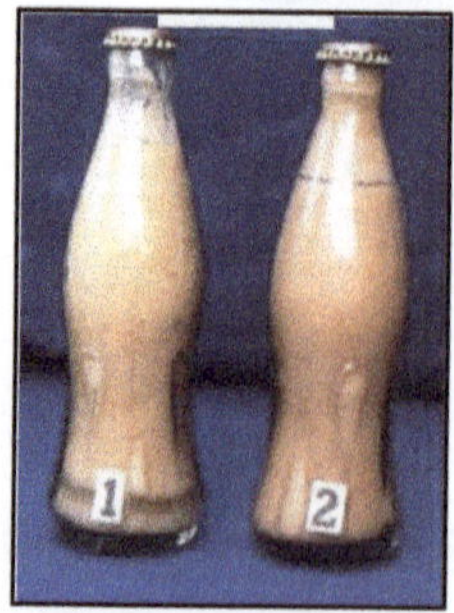

Narrow Necked Bottle with Metal Crown Cork for Liquid Products (Page No. 115)

Wide Mouthed Bottle with Metal Push on Lid (Page No. 115)

Wine Bottle with Natural Cork (Page No. 117)

Pepper Oleoresin in Narrow Mouthed Glass Bottles (Page No. 124)

"Standi Pack" with Wet Gravy Mix Processed by Heat Sterilization (Page No. 126)

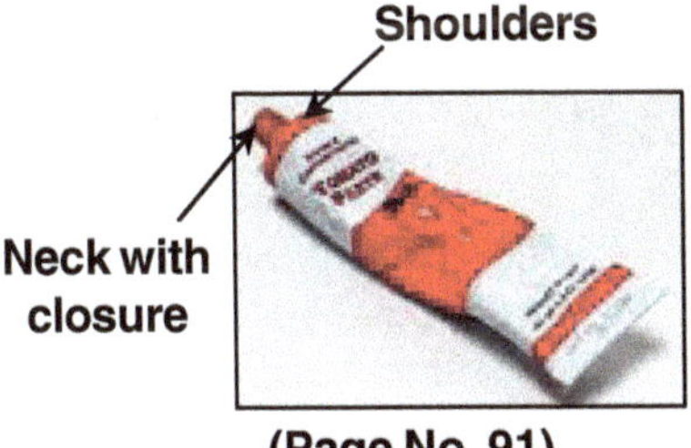

(Page No. 91)

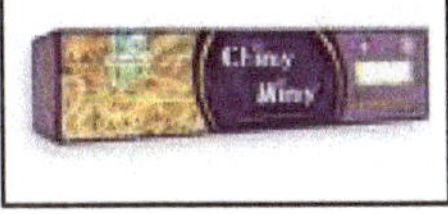

Vertically Filled Packages in Lamininated Polymeric Package (Page No. 132)

Aluminium Milk Cans of Different Capacities from 10 to 50 Litres for Short Distant Transportation of Fluid Milk and the Cans are of ISI Standards (IS1373) and (IS 4937) (Page No. 135)

Secondary Packaged Butter with Inner Parchment Paper and Outer Paper Chip Board Box (Page No. 137)

Ice-cream Cups made of High Impact Poly Styrene (HIPS) (Page No. 140)

Prawn in Tray with Overwrap and Blotters Underneath (Page No. 157)

Dressed Chicken in Vacuum Pack (Page No. 158)

Traditional Packing House of Fresh Fruits and Vegetables (Page No. 162)

Shrink Wrapped Meat in Tray (Page No. 157)

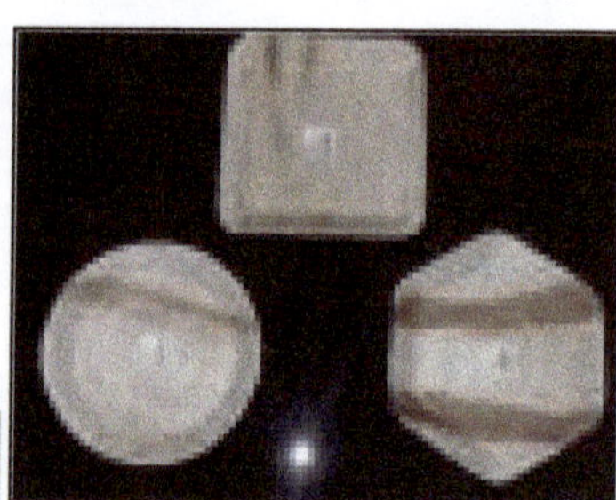

Areca Nut Leaves Sheath Trays (Page No. 170)

Bamboo Baskets (Page No. 170)

Plastic Crates (Page No. 174)

Jute Bags (Page No. 171)

www.ingramcontent.com/pod-product-compliance
Lightning Source LLC
Chambersburg PA
CBHW040756150726
48196CB00009B/577